Robert Collett

Bird Life in Arctic Norway

A popular brochure

Robert Collett

Bird Life in Arctic Norway
A popular brochure

ISBN/EAN: 9783337329457

Printed in Europe, USA, Canada, Australia, Japan

Cover: Foto ©berggeist007 / pixelio.de

More available books at **www.hansebooks.com**

BIRD LIFE
IN ARCTIC NORWAY

A Popular Brochure

BY

ROBERT COLLETT

Foreign Member, Z.S.
Professor of Zoology in the University of Christiania

TRANSLATED BY

ALFRED HENEAGE COCKS, M.A., F.Z.S.

London
R. H. PORTER
18, PRINCES STREET, CAVENDISH SQUARE, W

1894

TO

PROFESSOR ALFRED NEWTON,

M.A., F.R.S., F.Z.S.,

PROFESSOR OF ZOOLOGY AND COMPARATIVE ANATOMY IN
THE UNIVERSITY OF CAMBRIDGE,

THE ENGLISH EDITION

IS

(BY PERMISSION)

DEDICATED.

This paper was written for the Second International Ornithological Congress, held in Budapest, in May, 1891, and a portion of it was read during one of the meetings of the Congress. The Author wishes therefore to call attention to the fact that it is adapted to an audience whose conceptions of the natural features of the country described could not be supposed to be great. And in order that the character of the paper in its Norwegian form should be entirely popular, sundry sections or remarks contained in the original edition, were omitted, which might be assumed to have interest for specialists rather than for general readers.

Examples of all the species of birds treated of, may be found set up in the Christiania University Zoological Museum, the greater number of them collected in the very regions, whose nature is here sketched from observations during seven summers' wanderings in those parts of the country.

NOTE BY THE TRANSLATOR.

"Bird Life in Arctic Norway" was originally written for the congress in German, a revised and somewhat condensed edition being subsequently published in Norwegian.

It is this latter edition that, at Professor Collett's request, I have, and I fear very inadequately, translated. However imperfect the result, the attempt has given me much pleasure, as some slight acknowledgment of kindnesses shown me by the author, in my many migrations through Christiania.

The appendix was added at the suggestion of Mr. Sclater, by whose advice also I have (with the author's consent) altered the synonomy and arrangement of species to accord with the British Ornithologists' Union list, as more convenient for English readers.

I have to thank Mr. T. Southwell, F.Z.S., for kindly looking over the proof-sheets, and making many valuable suggestions.

A. H. C.

ERRATA.

P. ix., l. 8, *for* "as the Swiss Alps themselves," *read* "like the Swiss Alps."

P. x., second line of last paragraph, *for* "open" *read* "stretch."

P. 6, l. 14, *for* "which drowned," *read* "drowning."

P. 11, l. 6 of 3rd paragraph, *for* "fifteen" *read* "sixteen."

P. 11, l. 1 of 4th paragraph, *for* "extends" *read* "lies."

P. 14, l. 14, *for* "songsters" *read* "warblers." (The sentence should have been altered in accordance with the B.O.U. arrangement.)

P. 14, l. 8 from bottom, *dele the last* "the," and next line *for* "heads" *read* "head."

P. 17, line 4 from bottom, *for* "especially of the" *read* "especially the."

P. 27, last line but one, *for* "occupied" *read* "covered."

P. 36, last line, *for* "thick" *read* "turbid."

Appendix, p. iii., l. 11, *for* "13" *read* "12," and l. 13, *for* "40" *read* "41."

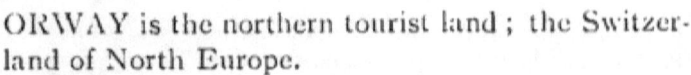

ORWAY is the northern tourist land; the Switzerland of North Europe.

Norway also has its shining Ice-braes, larger than any other glaciers in the continent of Europe. It has waterfalls of matchless beauty, which precipitate themselves with noise like thunder, a depth of 700 feet, to where no human eye can reach, nor foot tread their margin. It also has its Alps, mighty and snow-clad, as the Swiss Alps themselves, although of less considerable heights; but the nature and ordinary character of the Alpine scenery in the two lands, are entirely different. At the foot of the Swiss Alps winds a belt of chestnut and walnut trees, on their slopes flourish even vines, and the inhabitants grow wheat on the shores of charming lakes or in the warm valley bottoms, where the mulberry-tree, the fig-tree, and the maize also flourish. The Norwegian Alps rise as a rule from a lofty mountain plateau; their summits are sometimes decidedly imposing, and their confused masses of mountain-ridges and deep valleys picturesque; but no trees wreath the bases of the peaks, except the uppermost stragglers of the stunted pines and birches; and in the corries on the mountain-sides and upon the plateaux, large tracts are covered with thickets of the silver-gray mountain-willow, which in turn are succeeded higher up by monotonous areas covered by various species of lichen.

Here, in the belt of gray mountain-willow (*Salix lapponum*, and *S. glauca*), is the home of the Willow Grouse, here is the goal of the sportsman's longing; but the height—between three and four thousand feet—above the level of the sea, and the more northerly situation, cause these mountain-wastes from which our proudest peaks rise directly, to be only inhabited in the summer-time by scattered "Sæter" folk (or mountain-dairy people), or to be frequented by the numerous summer visitors,—Tourists and Sportsmen.

But Norway has another group of Alps, which Switzerland has not. This is the continuous mountain range which raises its snow-clad summits directly from the sea itself, in an almost unbroken chain from the borders of Nordland, straight on, nearly up to North Cape.

Through six degrees of latitude these sea Alps form a continuous wall, shutting off the inner, somewhat lower districts; and ending particularly picturesquely, when they form as in Lofoten, a separate branch, which bends out from the principal series, and extends like a row of gigantic shark's teeth, floating upon the surface of the sea, straight out into the Arctic ocean. "Lofoten's wall" is seen to the greatest perfection when it is viewed from the south, illuminated by the rays of an evening sun, until the last points of land lose themselves far away in the west, more than 60 miles* to seaward.

At Bodö, a little south of Lofoten, but also within the Arctic circle, begins Norway's Arctic region, the land of the midnight sun.

From here the course lies across fjords or through narrow sounds, where the mountains which rise to over 3,000 feet, are wreathed on the sea-ward side with a belt of dark-green mountain-birch (*Betula odorata, var. alpigena*); whilst their tops, the whole summer through, bear the remnants of their white winter dress. Near the North Cape this Nordland scenery first begins to lose something of its character: the mountains become lower, and North Cape itself forms a plateau with vertical cliffs down to the Arctic Ocean, the height of which hardly reaches 1000 feet.

On the eastern side of the Cape the deep fjords of Finmarken open towards the south, and Porsangerfjord, Laxefjord, Tanafjord, and lastly Varangerfjord all cut their way, one after the other, deep into the land. Here the coast scenery is quite different to what we have seen hitherto. The shores are lower, partly swampy, and clothed with vegetation, in some places reminding one of the Tundra region in their fauna and flora ; here is the home of the Arctic waders (*Tringa* and *Totanus* genera), which leaving their winter quarters in Mediterranean lands and Africa,

---

* 100 Kilometers.

select spots, often while the snow is still covering large tracts, where they may hatch out their young, during our short but sunny summer.

Finally, from the inmost heads of the fjords, the land rises up to the monotonous wastes of Finmarken, or Lapland proper, clothed with sparse birch-forest, and pierced by rivers and lakes, regions which have great attractions for the sportsman and naturalist, but less so for the ordinary tourist, who only seeks after diversified scenery, and has not patience enough to wage throughout the short polar summer, a semi-hopeless war against the mosquitoes which swarm here.

We shall in what follows, treat of some few traits of the Bird-life in that portion of our land. Let us therefore make in imagination a rapid flight to this north-westernmost corner of Europe, wander through the three natural zones, whereof Arctic Norway consists, and each of which shows quite peculiar characteristics, and in the meanwhile seize by degrees, whatever particularly strikes our attention on the way. The three natural divisions referred to are :

I.—The coast district and the belt of islands girding the coast up to North Cape.

II.—The deep fjords of the Arctic Ocean and the adjacent river basins in East Finmarken.

III.—The interior plateaux of Finmarken, or Lapland proper.

I.

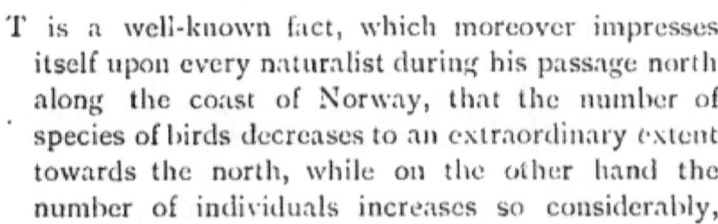T is a well-known fact, which moreover impresses itself upon every naturalist during his passage north along the coast of Norway, that the number of species of birds decreases to an extraordinary extent towards the north, while on the other hand the number of individuals increases so considerably, that hardly anywhere else in our continent do we see bird-life more richly displayed, than just when we stand on the furthest point of North Europe facing the Arctic Ocean.

It is especially in the great colonies, "The Bird Rocks," where this swarming bird-life exists.

Such bird rocks make their appearance at intervals along the whole of the coast of Norway from Stavanger, off and on, up to Varangerfjord and the Russian frontier, exactly as we know them on the coasts of Scotland and of the Faroe Isles. But while the bird-rocks of these districts, the west-European on the one side, and the Norwegian on the other, have of course, most of their breeding species in common, as for instance the Guillemot and Razorbill, the Puffin, the Cormorant and Shag, besides some gulls, especially the Kittiwake (or " Three-toed Gull," *Rissa tridactyla*): it is remarkable that the Norwegian bird-rocks wholly lack several species, which form to some extent their chief occupants in the west-European district.

This is not only the case with the more pelagic species, which belong to the more open parts of the Atlantic, as for instance the two Petrels (*Procellaria pelagica*, and *P. leucorrhoa*), the Gannet (*Sula bassana*), and the characteristic Shearwaters (*Puffinus anglorum* and *P. major*), which certainly occur occasionally on the Norwegian coast, but never breed there, although several of them dwell as near us as the Faroes; but also applies to wholly arctic forms, such as the Fulmar Petrel (*Fulmarus glacialis*), which likewise breed in multitudes on the Faroes and right down to St. Kilda, but never nest on the coast of Norway.

And just as the northern point of Norway forms the northernmost breeding-place for the Razor-bill (*Alca torda*), so there are found inversely in Spitzbergen and the large archipelagoes in the Arctic ocean, certain extreme-northern species, which never breed so far south as Norway. This applies to a species of Tystie—Mandt's Guillemot (*Uria mandti*), Brünnich's Guillemot (*Lomvia bruennichi*), the Little Auk (*Mergulus alle*), the white-winged Gulls (*Larus glaucus*, *L. leucopterus* and *Pagophila eburnea*), and others.

The question occurs to us, where do these enormous hosts of Little Auks, Brünnich's Guillemots, and arctic Gulls pass the winter? In hosts, whose numbers we can form no conception of, they breed on Spitzbergen and the islands in the north of Franz-Joseph land, as far towards the North Pole as human eyes have reached. It cannot be said that they take up their quarters during the winter on the shores of north or west Europe. No doubt our shores are visited in the winter by a certain number of Little Auks, to which may be added stray flocks of Glaucous Gulls (*Larus glaucus*) and King-Eiders (*Somateria spectabilis*), or a solitary Ivory Gull (*Pagophila eburnea*) or Brünnich's Guillemot; but the Spitzbergen Tystie or Mandt's Guillemot (*Uria mandti*) is quite unknown on the European coasts. Here therefore is not their winter home.

When the polar night with its darkness of some months' duration broods over the Arctic archipelagoes, there are not many birds which are able to sustain life there. The Snowy Owl (*Nyctea scandiaca*), the Spitzbergen Ptarmigan (*Lagopus hyperboreus*), and perhaps one or two other species, do so; but the sea-birds forsake their nesting places, disappear out towards the open Arctic Ocean, and quest away where no human being has yet been able to follow them.

To the north of the arctic circle the bird rocks are more requent, and larger, than they are further south. With greater or lesser interval they extend in an irregular series from the outmost extremities of the Lofotens—Værö and Röst, up along the coast of Tromsö to Fuglö and Loppen; thence they make their appearance to the eastward in Stappen near North Cape; in Sværholtklubben in Porsanger; and in several crags on the Varanger peninsula to Hornö by Vardö; and on the south side of Varangerfjord near the Russian frontier.

What an inexhaustible field for observations, these bird rocks would be, if we could only pay longer visits to them.

It is well-known that among the rock-fowl both the Razor-bill (*Alca torda*), and the common Guillemot (*Lomvia troile*)* lay their single,—and in proportion to the size of the bird—colossal egg, on a projecting ledge in the precipitous face of the cliff, either quite in the open, or under a projecting slab of stone. The Puffin (*Fratercula arctica*) on the contrary, digs with its sharp claws, a long horizontal passage in the soft stratum of earth on the slopes of the cliff, between the luxuriant clumps of *Cochlearia* (scurvy-grass) and other coast plants: showing a considerable difference in the choice of a nesting-place between two so closely allied forms.

But the difference is even greater, if one looks at the young which issue from these eggs. The Razor-bill produces an almost entirely naked thing, which, when the mother is away from it, is obliged to balance itself, as best it can, upon the narrow ledge of cliff, exposed to the icy north wind, and frequently drenched by a snow squall or a cold rain. From the Puffin's egg, which lies some yard-and-a-half deep down in the close and sheltered tunnel, there comes a chick, clad in a downy covering so loose and fluffy, that it resembles a living ball of down, from which a beak and feet project.

The reason for this difference between the young of the two species is hard to understand. It is one of the many unanswered questions, to which the economy of nature can give rise so abundantly.

Where the space is scanty, the Razor-bills and Guillemots

* These two birds are respectively known in Norwegian as the Broad-billed-, and the Pointed-billed-, Alke; the latter is also called the Lomvi.— *Transl.*

sometimes lay,—which the two species commonly do in company,
—their solitary egg so close to one another, that the sitting birds

YOUNG RAZORBILL.

nearly touch each other. When then, one of them leaves its egg,
it frequently rolls out of position on the hard rock floor, and it

may often happen that two of them exchange places, and when there are as many as twenty birds sitting on the same shelf, it is not always possible for each individual bird to continue to sit upon its own egg. A certain degree of fellowship also prevails between neighbours from the fact that the male bird does not feel himself strictly bound to the one female alone.

Of the numerous bird rocks, I will only in a few words, touch on a single one, which in several respects is extremely remarkable;

YOUNG PUFFIN.

this is Sværholtklubben, situated on a projecting point a little to the east of North Cape. The fact is that it is probably the biggest in Norway, perhaps in the whole world.

Sværholtklubben is inhabited almost exclusively by a single species, a little gull,—namely the Kittiwake (or Three-toed Gull, *Rissa tridactyla*). When the tourist- or mail-steamer, full of passengers, approaches the steep cliff wall, whose height is about 300 feet, and whose breadth is considerably more, but where nevertheless, pretty well every available space of so much as a foot in size, is taken advantage of by the gulls, so that the whole wall seems white with the masses of birds resting on it, whilst at the same time "the air is darkened," as the saying is,—by the swarms on the wing. A shot is generally fired from the ship to cause the birds perched on the rock to take wing, but so accustomed are they to this attention, that it is only an inconsiderable portion which can be prevailed on to rise. When King Oscar II. visited Finmarken in 1873 in a man of war, the ordinary salute was first tried, but without particular success: then one of the corvette's large cannon opened its mouth: it resounded on the mountain wall like a thunder clap, which drowned the noise produced by the innumerable screaming throats, and then, says Professor Friis,* who was present, even the old individuals were forced to turn out.

Each spring, about the middle of May, when the eggs are laid, the proprietor of the Klubbe takes as many of them as he can reach with a long pole from the foot of the mountain; ropes are not employed here, as on the Faroe islands. The maximum clutch is three eggs to a nest; the yield is about 5,000 eggs, some years not so many; in others as much as double the number. These represent nearly 2,000 pairs of birds; all the remaining portion of the mountain-wall remains untouched. But moreover it appears, that for every breeding pair (with entirely white head) the cliff is inhabited by perhaps eight to ten young individuals, recognizable by the black ring on the nape of the neck, which do not breed. In determining the total number we shall reach up to millions, and these masses are crowded together like white snowflakes upon the narrow resting-places, and at the foot, of a single vertical, black area of comparatively insignificant extent.

On what do these swarms live? In greater or lesser companies, often in rows as straight as a line, train after train come in at fixed times of the day, passing from the sea or over the

---

* Professor J. A. Friis is Professor of Lappish in Christiania University, a well-known sportsman, and author of numerous interesting and valuable works.—Transl.

more open fjords towards their home, all crammed with food. This consists partly of fry of fish and their ova, especially of the Coal-fish, and partly of small crustaceans, which the sea-currents drive backwards and forwards, in enormous masses, both close to shore and out at sea. Among these are the little copepod *Calanus finmarchicus*, transparent as water; this food they share with the Rudolphi's Rorqual (or " Coal-fish Whale," *Balænoptera borealis*), a species of whale of medium size, of which in some seasons (as in 1885), nearly 800 head are captured on the coast of Norway. Another crustacean occurring in large quantities, is the " Kril" (*Euphausia inermis*), a small species of Thysanopod, half-an-inch long, which also forms the chief food of the Blue Whale (*Balænoptera sibbaldii*), the largest of all now living (or probably of formerly existing) creatures, when this sea giant stands in under the Norwegian coast in the summer months. And the Blue Whale in turn must yield its life to the explosive harpoon-shell, whilst it lies on the surface of the water and gorges on these small crustaceans, which form its only food during its sojourn with us, and of which even up to ten barrels* have been found in the capacious stomach of some of the examples that have been captured.

The Kittiwake builds its nest upon the narrow shelves on the steep cliff wall, where they hang like swallows' nests out over the breakers. In the course of years they become constantly larger, since they are extended and added to each year, so that at last they may reach a height of several feet. They are a tangled mass of straw (bents) and sea-weed, copiously saturated by the droppings of the birds and their young ; sometimes they are situated so low down, that the spray of the surf reaches them, but such things affect neither the young nor the sitting birds.

In the winter the mountain remains deserted. The birds, which are often seen in whole clouds at sea, or under the land, engaged in fishing, never then sit on the Klubbe. In March they suddenly return, and occupy it all at once; at the end of August, when all the young have taken to the water, they once more leave it.

* A Norwegian *Tønde* (= barrel) holds about 30 gallons (= 3.83 bushels).— *Transl.*

YOUNG KITTIWAKES.

Besides the Kittiwake, there breed on Sværholtklubben only a few stray Razorbills and Guillemots (*Alca torda* and *Lomvia troile*). some Tysties (*Uria grylle*), and the two kinds of Cormorant (*Phalacrocorax carbo* and *Ph. graculus*), but all these together are quite lost sight of among the countless gulls.

High over these swarms of birds there commonly soar in majestic calm, a pair of Sea-eagles (*Haliaëtus albicilla*), which have placed their inaccessible nest up under the summit of the mountain. But the constant sight of the mighty pirate, who by merely making a swing with his wings can at any time possess himself of one of the young Kittiwakes from the open nests, has accustomed the denizens of the cliff to the danger, and they take little notice of him. Should, on the contrary, a Gyr-falcon (*Hierofalco gyrfalco*) on its piratical expedition come too near the colony, the effect is quite different; the exasperation of the inhabitants is aroused, and as it is not a member of the community, it is pursued with loud screams, so long as it remains in sight.

Before we take leave of the bird-rocks and their residents, we will record an event of historic interest, which is connected with one of the most easterly of the bird-rocks of Finmarken, namely, Hornö by Vardö. In the year 1848, there was shot here by a man still living, Herr L. Brodtkorb, a bird, which though it was not preserved for posterity, was yet so fully described by the gunner as to be immediately recognized by naturalists, and the remembrance of it also has been so faithfully preserved, that any mistake is hardly conceivable. Then, indeed, in all probability, was shot the last existing example of that remarkable bird, the Garefowl, or Great Auk (*Alca impennis*). As the author has already elsewhere* referred more fully to this occurrence, and as the Garefowl never seems to have been a constant inhabitant of the coast of Norway, we will not here detain ourselves longer with its history.

We must briefly mention what species of birds may be considered as the most characteristic in the belt of islands which gird the coast of Norway.

* *Mitth. d. Ornith.* Vereines in Wien, 1884.

Along the whole coast-line there occurs hardly a holm, or an island large enough to provide sufficient food to keep a couple of sheep during the summer, without its being also inhabited by a pair of Oyster-catchers (*Hæmatopus ostralegus*), some Ringed Plovers (*Ægialitis hiaticula*), often also by the Turnstone (*Strepsilas interpres*), and some small Gulls and Terns (*Sterna macrura*). If the island is bigger, and more covered with heather or grass, there may nearly always be found there, in addition, one or two pairs of Eider Duck, and a sprinkling of the larger species of Gull, especially Herring Gulls (*Larus argentatus*).

As we approach the larger bird-islands or egg-holms the bird-hosts are recognizable from a long distance.

Upon these egg-holms is the home of the Eider. Everywhere among the heather or in the scrub, the ducks are sitting close upon the five large yellowish-gray eggs, surrounded by the fluffy wall of down ; it is well-known that they often place their nests quite close to the houses of the human inhabitants, even under the door-steps, or the floor of the kitchen. The Eider is often the islanders' only domestic fowl ; through the whole summer the broods lie scattered along the shore-edge, and the little brown-black ducklings dive gallantly into the surf after mussels, and other small creatures, and they also eagerly search for the fish-offal that is thrown away. Complete harmony prevails mutually between the families ; if the young get separated from their mother, they attach themselves to the nearest duck that they meet with, and one sees not unfrequently a duck in this way at the head of a row of over twenty small ducklings, which follow her like a string of beads on the surface of the water.

The Greater Black-backed Gull (*Larus marinus*), the Grey-Lag Goose (*Anser cinereus*), and the Arctic or Richardson's Skua (*Stercorarius crepidatus*) are also among the most frequent of the inhabitants of the egg-holm. And upon the largest of them, where the protection is strict, it not unfrequently happens that quite strange species settle themselves down to nest. Thus on B—— in Lofoten there have bred for many years a pair of Barnacle Geese (*Bernicla leucopsis*), a bird which nests nowhere else in the country.

The nearest relation of the last-named species is the Brent (*Bernicla brenta*). This goose never nests on the continent of Europe, although the young birds now and then spend the summer with us, and it is a well-known visitor to the country at migration times.

In the spring Brent Geese push in under the Naze (Lindesnæs) on a fixed day, towards the end of May, in large skeins, and more follow on the succeeding days; in rows as straight as a line they sweep compactly over the surface of the sea along the whole coast until they reach the outermost north-westerly skerries. Then the crowds sweep further afield, so as to fetch their nesting-places in Spitzbergen and Novaya Zemlya; and the sealers and Arctic travellers who have stood upon the northern point of Spitzbergen have seen them wandering yet further over the snowy sea, seeking still more northerly archipelagoes, which as yet no human being has trodden.

Tromsö offers a good opportunity for observing bird-life as it exists on the islands which fringe the Arctic coast, being one of the lower and more thickly inhabited of them, situated nearly in the 70th deg., N. Lat., and bearing Finmarken's capital city of the same name. In that town there exists a veritable Arctic Museum, whose industrious scientists have for fifteen years published their "Tromsö Museum's Annual," the most northerly scientific journal in the world.

This pretty little island, which extends like a green knoll within a circle of snow-clad mountains, is thickly over-grown with birch-wood, alternating with tracts of marsh and some cultivated fields. Although the former richly varied animal. life has of late years somewhat decreased on account of the in-creased area under cultivation, still a stroll along the gardens,* with which nearly every house outside the town is provided, and in the surrounding birch-groves, will in a short time make us acquainted with several very characteristic birds, which have their place of summer resort here.

* Where in particular the stately *Heracleum panaces* with its unusual luxuriance excites the traveller's admiration.

Everywhere is to be heard the wild cry of the Fieldfare (*Turdus pilaris*),—that bird so characteristic of our sub-alpine regions,—which nests everywhere, particularly in the coast districts, in large colonies ; nor is it anywhere wanting in the thick birch woods, which as a rule clothe the sides of all the mountains in the Arctic district. Here are found nest after nest, more or less close to each other, but never more than one on each tree : where trees are wanting, or where the birds are undisturbed, they take up their quarters upon the verandahs of the houses, and during the nesting season, are somewhat obtrusive with their shrieking and noise.

In each colony of this kind there are generally found established one or two pairs of Redwings (*Turdus iliacus*), which are even more obtrusive than their larger cousins. Other small birds also commonly settle down in the midst of the colony, because they well know, that the stout-hearted thrushes will keep all sorts of robbers away, or, at any rate, give them timely notice of the danger.

In every meadow we commonly find a pair of Blue-headed Yellow Wagtails (*Motacilla flava*). This delicately made bird, with its lemon-coloured belly, a near relation of our common White Wagtail (*M. alba*) inhabits in Norway almost exclusively the higher-lying regions, such as the Sæter (= mountain dairy) inclosures in the southern high Fjelds, or grassy-bottomed spots in the arctic parts.

In this northern race the hood of the male is dark blue-gray or nearly gray-black (*M. cinereocapilla*, Savi), whilst in the typical form in the lowlands of mid-Europe, it is ash-coloured with greenish intermixture.

From the thickets in the gardens, and from the willow bushes in the moister places, may be heard the best songster of the arctic district, the Blue-throated Warbler (*Cyanecula suecica*), delivering its weak but harmonious song ; certain notes in which remind one of the sound of a distant bell, and the peasants call it therefore in several places the Little-bell bird. In Norway it is entirely Alpine in its habitat, and never breeds in the lowlands.

Another remarkable songster, which inhabits the willow

thickets of Tromsö in very large numbers, is the Sedge-warbler[*] (*Acrocephalus phragmitis*), the only species of its genus, which occurs in Norway. And with us it inhabits almost exclusively the regions north of the arctic circle, whilst everywhere else in Europe it is common in all lowland reed beds.

Like most of the genuine migrants of the order *Passeres*, the small birds above alluded to, do not reach these their northern resting-places across the southern parts of Norway, but by the eastern route, across Russia and the Baltic provinces. One consequence of this is, that several species which appear frequently in Finmarken, such as the Red-throated Pipit (*Anthus cervinus*), and the Siberian Willow-Warbler (*Phylloscopus borealis*), are either not met with at all during migration in the southern parts of the country, or appear there only occasionally and accidentally.

Particularly numerous in the birch-woods of Tromsö are the Brambling (*Fringilla montifringilla*), and the Mealy Redpoll (*Linota linaria*); and the males, ubiquitous and irrepressible, are to be heard in the early summer practising their best arts of song, although the long, harsh call-note of the first mentioned,—the only one which it is capable of producing,—hardly deserves the name of song.

But of all the numerous bird-voices, which meet us as we ramble on a spring day through the still leafless groves of Tromsö, there is none more surprising, than the guffaw of the male Willow Grouse (*Lagopus albus*). This vivacious bird, newly arrayed in its handsomest spring plumage, with the dark chestnut-brown head and neck contrasting sharply with the remainder of the still snow-white dress, never omits to croak, when it settles, after being flushed. Its voice is curious, and indescribable, and is a characteristic feature seldom lacking in the life of our mountain birch-woods.

A fortnight later the hen lays her eggs among the heather or under a birch bush, often close by the frequented roads of the island, or even within the garden or enclosure of the villas. But she lies untouched and safe; her brown-speckled back matches

---

* It may be worth noting, to avoid any chance of confusion, that the Norwegian name is Rörsanger, which means literally Reed-warbler; a species which does not occur so far to the north as even the south of Norway.— *Transl.*

the surroundings so perfectly that it is almost impossible to be certain at the distance of a few paces, whether one still knows exactly where she is.

Upon the open bogs there breed a motley collection of arctic or sub-alpine birds, mingled with coast forms and more southern species. Here occur the Red-throated Pipit (*Anthus cervinus*), the Common Snipe (*Gallinago cælestis*), the Great Snipe (*G. major*), Temminck's Stint (*Tringa temmincki*), Curlew and Whimbrel, several Sandpipers (*Totanus glareola, T. canescens*, and *T. calidris*), the Red-necked Phalarope* (*Phalaropus hyperboreus*), and various "broad-billed" or "gray" Ducks (*Anas, Mareca, Dafila, &c.*).

And among the common European small-bird fauna, we shall also find various true songsters resident here, as the Redstart (*Ruticilla phœnicurus*), Hedge-sparrow (*Accentor modularis*), the two Fly-catchers, the White Wagtail, and many others. The White Wagtail follows mankind right up to the most northerly point of the land inhabited—the fishing-station on Gjæsvær by North Cape.

Hither come, as more or less accidental visitants, the Starling, the Swallow, the Sky-Lark, the Woodcock, the Land-rail, and also the Quail, besides several others.

In the winter the birch woods of Tromsö remain nearly deserted, and at that time, besides the Willow Grouse, one meets with only an occasional Scandinavian- or Northern-Marsh Tit (*Parus borealis*), and the Lesser Spotted Woodpecker (*Dendrocopus minor*), some Yellowhammers, Mealy Redpolls, Bullfinches, a solitary Tree-creeper (*Certhia familiaris*), some Golden-crested Wrens (*Regulus cristatus*), and also rarely—the Longtailed Tit (*Acredula caudata*), this last being of course the type-race with the entirely white heads the only form which occurs in Norway. (The young in nestling plumage have, however, dark eye-stripes.)

Lastly, we must mention among the stationary species, the three well-known relations—the Magpie, the Hooded Crow, and the Raven. All these go up as far to the north as men and food are to be found, and the Magpie builds its large nest on the faggot-piles by the houses even on Gjæsvær, directly by the

---

* Norwegian, Svømmesneppe—Swimming Sandpiper.

North Cape. The Raven is everywhere along the coast deservedly regarded as a harmful bird, because out of wanton mischief it pecks in such a way at the fish hung up to dry, that the wooden frame on which they hang falls down, and the fish lie on the ground and are spoilt. Up here it is common everywhere, and one may often in the autumn meet with flocks of twenty to fifty head collected together.

II.

THE steamer has passed the point of North Cape, and anchored in Hornvigen, a little bight on the Eastern side of the mountain, to allow the tourists the opportunity of ascending the northernmost outpost of Europe against the Arctic Ocean. We go ashore, wade through plants, as high as a man, of the vigorous *Mulgedium alpinum*, *Archangelica officinalis* (" Kvanne " = Angelica), and an extremely luxuriant form of Scurvy-Grass (*Cochlearia officinalis*), which reaches above our knees ; we clamber up along the narrow cleft of the mountain in order to reach the plateau proper, which forms the termination of the Cape, before it topples over sheer down into the Arctic Ocean. We post ourselves on the slope facing the Arctic Ocean ; the time is between one and two at night ; the sun's red disk stands high above the horizon ; the sea lies burnished like a looking-glass, and one seems to be able to see right up to the North Pole ; but we cannot linger long—the ship is waiting for us down below.

Still, we get just enough time to observe that even this desolate plateau has bird inhabitants. Besides the ubiquitous Wheatear—which is not wanting anywhere in our country, from the naked rocks of the coast and up on the mountains to the snow-line—we here come across a pair of Ringed Plovers (*Ægialitis hiaticula*), which run away, crying anxiously, among the small stones which cover the plateau like a floor ; and, if we are lucky, we may discover the four down-covered young, which, like tiny grayish-brown lumps of down, lie flat among the gravel, where they remain motionless so long as the danger lasts.

Should we have time for a longer excursion, we may,—in the

warm hollows, that indent the slopes of North Cape towards the south,—find a couple of well-known species of birds, which attend us in all parts of the land. One of these is the Cuckoo, which in these regions entrusts its eggs to the Pipits and the Wheatear; the other is the Willow-Warbler (*Phylloscopus trochilus*), which, in the highest willow thicket that even here manages to sustain life, executes its somewhat tedious song as indefatigably as it does in the beech-woods of Central Europe. Its nest, as round as a ball, and which in these storm-vexed parts is large and fluffy, is lined with a handful of the white winter-feathers of Willow Grouse. Its relation, the Chiffchaff (*Phylloscopus rufus*), on the contrary, hardly goes further north than Saltdalen, only a little way beyond the Arctic circle.

We quit the plateau again, and go slowly down through the steep cleft in the mountain, where the path winds among loose stones and snow-fields. From high up the precipice are heard the melancholy notes of a solitary bird, which sound almost like the cry of a young bird separated from its mother. That is the Ring Ouzel (*Turdus torquatus*), which sings to his mate, whilst she is sitting on her brown-speckled eggs under a tussock of grass up on the mountain, or is engaged in rearing her young.

The most emphatically Arctic representative in the group of small birds, is the Snow Bunting (*Plectrophanes nivalis*). On the precipice of North Cape, and on the stacks furthest out to sea, upon the plateaux in the interior, and upon the archipelagoes of the Arctic Ocean, everywhere one finds scattered pairs of these birds established, whose simple summer plumage, composed of pure black and pure white, harmonises so remarkably with the ground that they have chosen to inhabit. From one of the dark boulders, alternating with the snow-drifts which the short summer is unable to thaw, or from the highest point of such a snow-field, the male during the nesting season sings his pretty and variable song, sounding quite cheerful in the dreary surroundings; then he flies down among the stones, and comes back shortly with his beak full of insects, especially of the large Cranefly (*Tipula*), to feed his sitting mate or the young under the slab of stone.

We have again reached the foot of the Cape, where our

friend, the Conchologist, has seized the opportunity to hunt out from under the stones and among the tussocks of flowering *Silene acaulis*, *Cerastium alpinum*, *Rhodiola*, and *Viscaria alpina*, a somewhat varied Fauna of Insects and Land-Molluscs,* which this soil, only free from snow for three short months in the year, is yet able to produce.

And by the edge of the sea, among the large stones, we are greeted at our departure by a pair of Rock-Pipits (*Anthus obscurus*), the only species of its family (which includes the Wagtails), which passes the winter with us.

We leave North Cape behind us and find ourselves in Porsanger Fjord, the first of the large Fjords of the Arctic Ocean, which cuts inwards to a depth of eighteen geographical miles into the mainland of Finmarken. The shores of all these Fjords resemble each other, and the same characteristics repeat themselves more or less in them all. The mountains here are lower, most frequently naked and rounded; the coast is often level and flat, and as a rule clothed with vegetation; either with ling, alternating with willow scrub and tracts of swamp, or where there are permanent inhabitants, interspersed with small green plots of meadow.

But on these heather-clad and boggy shores, and in the bottoms of the valleys, which in nearly every place are clothed with vigorous birch-woods and the most luxuriant growth of grass, there are spots where the naturalist will find a bird and insect fauna so rich and peculiar that it can be exceeded in few other places in the country.

Such a place is S—— T——. The island is only a Norwegian square mile in extent,† and quite low; its surface, rubbed smooth in the glacial epoch, and very slightly undulated, is treeless, but covered with a thick layer of peat, overgrown with short plants and heather, among which, here and there, gleams the reflection of water.

* Such as *Conulus fulvus*, *Vitrina angelica*, *Arion subfuscus*, and *Alæa arctica*.
† About 7½ English miles. — *Transl.*

Besides the large Coal-fish fishery, which takes place here in the summer, the island has economically speaking, two glories. If one lands here on a beautiful spring-like day in June, almost the entire surface will be white with the large corollæ of the cloud-berry (*Rubus chamæmorus*).* And when autumn comes, these millions of plants will bear their large yellowish-red berries, which can hardly anywhere else reach a more luxuriant development than in favourable years they do here.

The second glory of the island is its supply of eggs and down. This island is the nesting-place of one of the largest colonies of Eider Ducks in the country; and as soon as the ducks have settled under the small tussocks of ling, or between the crevices in the layers of peat, the valuable down is stripped (as a rule only once) from each nest; and when the "down-harvest," is complete, the proceeds fill an entire room up to a man's height. As the island belongs to the jurisdiction of the chief magistrate† of Finmarken, the lessee, who is its only inhabitant, must deliver annually two casks‡ of cloud-berries and 48 kilograms § of thoroughly-cleaned down, out of the produce.

The island is therefore strictly watched during the breeding-season, for there are many poachers among the more vagrant fishermen and Laplanders, and no one is allowed to set foot on it before the young Eiders have taken to the sea.

Numberless gulls, belonging both to the black-backed (*Larus marinus* and *L. fuscus*), and to the blue-backed species (*L. argentatus* and *L. canus*), nest in colonies all over the island, each species commonly occupying a space to itself, in which none of the others occur. Hard by there breed a considerable number of Grey-lag Geese (*Anser cinereus*), but the two remaining species of Wild Geese (the Bean Goose, *A. segetum*, and the Lesser White-fronted Goose, *A. erythropus*), belong to the inland parts of the country. These gulls and geese supply the tenant with many thousands of eggs annually.

Among all these swarms of birds there breed numerous other species, which enjoy the advantage of the quiet which prevails

* It is impossible for anyone unacquainted with Norway to understand the extreme appreciation in which these berries are there held.—*Transl.*

† Amtmand.   ‡ Tönder, see foot-note, p. 7.   § 3 qrs. 21 lbs. 13½ oz.

on the island; and though no raptorial bird nests here, yet many of the gulls are veritable birds of prey, and sundry of the newly-hatched Eider Ducklings are snapped up, on their way from the nest to the sea, by the voracious Greater Black-backed Gulls (*Larus marinus*), and disappear in a twinkling in their capacious gullet. Even the Cloud-berries are devoured with avidity both by gulls and by others of the feathered inhabitants of the island.

The breeding season is over; it is in the month of July, and we set out upon our ramble over the island. Here and there on the small peat hillocks one hears the melancholy call-note of the Lapland Bunting (*Calcarius lapponicus*), and the male in his bright coloured summer plumage, with the broad reddish-brown collar and the yellow beak, regards attentively and without the least shyness, the intruder who sets foot in his domain. From the more swampy places sounds incessantly the song of the Pipits, when the male mounts a certain distance into the air, and sinks again on to a tussock or stone. Of the Pipits, the Red-throated species of the eastern arctic regions (*Anthus cervinus*) is, upon this island, nearly as common as the Meadow-pipit; and as it flies, the difference between it and the latter species is quickly seen; it appears larger, and has a longer and sharper call-note. The two species are commonly found intermingled in the same locality; the nests and eggs show no constant difference, though the eggs of *A. cervinus* are generally furnished with spiral lines, which as a rule are wanting in those of the other.

Where the surface is bare, and the ground stony or lichen-covered, one generally finds a pair of Shore Larks (*Otocorys alpestris*), established; this is likewise a Siberian species, which in comparatively recent times has immigrated from the East, and is now included among the more common of the small birds of Finmarken. Even before the snow has yet entirely melted, the hen is sitting on her eggs, in the nest lined with yellowish-white willow down; and the young, which are covered with an unusually soft and fluffy plumage, must often find themselves in a snow storm or sleet squall, which makes the ground perfectly white round the nest. During the time of incubation the male executes his insignificant song, whilst he, like a true Lark, flies round in circles for a long time, so high up in the air, that he sometimes is lost to sight.

If we approach the small pools of water, which occur here and there in the heathery spots, or upon the more grassy places, we shall soon pull up near a pair of small waders, which are fairly common here, and whose life-history and habits are well worth our attention.

Already, before we go ashore, we have noticed a wisp of small Sandpipers, which unlike all their relations, do not keep on land, but swim about near the shore, diving with their little head and slender beak, and constantly bringing up from beneath the surface some crustacean or some other lower form of life which is invisible to us. This is the Red-necked Phalarope (*Phalaropus hyperboreus*), which with the slender form of body of the Sandpiper, has acquired as close a covering of feathers as a duck, and for whom the water is the more proper element. Often they may be found in flocks on the sea far from land, rocking upon the surface in the strongest swell, like small specks of foam. But in the small tarns up in the interior of the country, between the leaves of *Comarum*, *Menyanthes* and *Hippuris*, we may be able to find their nest, or meet with the four delicately-formed brownish yellow young in down, being conducted by one of the parents among the water plants.

This one of the parents is, as is well known to the majority of my readers, always the male. With the utmost indifference to danger, he runs, anxiously screaming, directly in front of our feet, to divert our attention from the eggs or the small young. The females, on the contrary, which are a little larger, and purer coloured, keep themselves to themselves during the nesting season, and generally form the little flocks which we have seen floating about on the small pools of water, or in the sea close to shore, far removed from the burden of family life. Here the plainly-coloured male is the weaker sex, which must wholly and entirely undertake the hatching of the eggs and the bringing up of the young. This trait is by no means peculiar to this species, although hardly in any other case to so great an extent as the present, where it also takes expression in the colour of the male. It is more or less conspicuous with most of our arctic waders of the Stint and Sandpiper families. Thus it is almost always the male which sits, and leads the young brood, in *Tringa striata* (the Purple Sandpiper, Norwegian "Fjærepist" = "Ebb-piper") and *T. temmincki*, as well as in the Greenshank (*Totanus canescens*), the Wood Sandpiper (*T. glareola*),* and others; and if both parents are present, the male is always the bolder, the female more cautious, and also in better condition, than her mate.

In company with *Phalaropus* there commonly live a few pairs of Temminck's Stint (*Tringa temmincki*). This northern species, hardly larger than a sparrow, is numerous in heather-covered localities in the arctic parts of Norway, and here generally nests in small colonies on low-lying tracts overgrown with willows and *Empetrum*, not far from the shore, sometimes even in the middle of the small plots of meadow by the Laplanders' huts, if only there are small pools of water in the neighbourhood where they can search for food. At these regular feeding-places they put in an appearance several times a day during the nesting season to look for food.

During the whole time of laying, the male of this little sand-piper performs a peculiar play, consisting of flying exercises, combined with song, all to amuse his mate during the first period of their wedded life. With quivering wings he mounts, almost like a lark, singing and twittering, up in the air. Here he flies about, in a circle, at an inconsiderable height, during a trilling passage of the song, and sinks at last with raised wings, still singing, down upon a stone or on the top of a bush. The common Dunlin (*Tringa alpina*) also has a similar but far less elaborate play, and it is also known in the Knot (*Tringa canutus*).

There still remains to be mentioned, among the island's small waders, the smallest of them all, *Tringa minuta*. This Little Stint of the far north, it is true, appears during the migration periods, sometimes even in large numbers, alike in the south of Norway and on the other coasts of Western Europe, but about its summer haunts, and its breeding history, there

---

* The Norwegian name of this species is Grönbenet Sneppe = Green-legged Sandpiper, while that of the Greenshank is Glutsneppe = a corruption of *Glottis*-Sandpiper. — *Transl.*

were, even up to within the last few years, only published a couple of short notes made by Middendorff during forty years on the Taimyr peninsula in North Siberia (about 74° N. lat.).*

Here in Norway it seems, however, on the whole to nest only occasionally. It generally elects to take up its abode in companies of from one to some few pairs, in the middle of colonies of its nearest relation, *T. temmincki*, and its behaviour during the nesting season is almost exactly like theirs. It seeks its food at the same feeding-places by the swamps on the coast ; and by its extremely anxious demeanour, when one approaches its nest, it discloses as artlessly as its companions, where the four prettily-pencilled eggs, so highly valued by oologists, lie amongst the heather. The nest is even more skilfully constructed than its relation's, and is lined at the bottom with a thick layer of fine grass bents, almost like that of a pipit.

During the nesting season this bird also performs a play, which is executed much in the same way as with the preceding species ; but besides that, the male, and sometimes the female, utters a delicate twittering song from the ground in proximity to the nest. Here also it is the male on whom the burden of rearing the young is essentially incumbent.

We pause at length by the largest of the pools of water, which glitter among the cloudberry flowers and the layers of turf. Numberless gulls make their toilet here, as they prefer for this purpose the fresh water to the sea. Pair after pair of the Red-throated Diver (*Colymbus septentrionalis*) lie scattered over the water, or exercise their two tiny brownish-black young in diving in the small pools of water in the neighbouring swamp. On this island, where they are protected, this species, otherwise so little gregarious during the breeding season, has joined together into small colonies, a thing which has hardly ever been observed elsewhere : and when the author, in 1872, visited the island for the first time, there were at least thirty pairs nesting in its different tarns. All these pools, whose water is black and mixed

---

* In June, 1872, on this very island, the author met with several pairs of this species settled, and under circumstances which plainly showed that they bred here. But in 1875, as is known, Seebohm and Harvie Brown first found both eggs and young in the Gebet Tundra, near the mouth of the Petchora. In 1880 the author found several nests in two different localities on the shores of Porsanger Fjord.

up with mud and guano, are entirely destitute of fish; but a peculiar crustacean, belonging to the group of Phyllopods (*Polyartemia forcipata*), occurs here in large numbers.

Its nearest relation, the Black-throated Diver (*C. arcticus*), occurs but rarely on the low islands on the coast, but nests by preference in the inland lakes in the interior of the country; and the Great Northern Diver (*C. glacialis*), the largest of the three, whereof young, non-breeding individuals are met with in summer on our southern coasts, perhaps never nests in Norway.

It is also worth mentioning that the western-arctic *Colymbus adamsi*, one of the largest web-footed birds of the North Atlantic, and whose proper habitat must be said to be but little known, has of late years been brought into our museums several times, especially in autumn, from the coasts of Norway; in most in-stances from the west and south coasts of the country, though some have also been met with on the coast of Finmarken.

Before we quite leave T—— and the northern coast regions of the country, it will perhaps be of interest to point out which species of arctic waders and web-footed birds are never found nesting within the boundaries of the country, but appear in large numbers during the periods of migration, or at other times of the year.

It is well known, in what large flocks many of the arctic species of *Tringa* annually appear in the autumn on the west coast of Europe. By the middle of August there arrive upon our coasts the young of the year of *Tringa canutus* (the Knot, sometimes called by us the Iceland Sandpiper, or in Jæderen* "Grel"), and of *Tringa subarquata* (the Curlew Sandpiper, or in Norwegian, the Curved-beaked Stint), in company with Sanderlings (*Calidris arenaria*) and others. But whence come these hosts in thousands and thousands, which gradually disperse themselves over the coasts of the North Sea lands? We only know that they come from the north, and the exceedingly few cases, in which the eggs

* Jæderen is nearly the S.W. most corner of Norway, the district between Stavanger and Egersund.—*Transl.*

or young of these species have been found with certainty in the extreme north, lead to the conclusion that their chief nesting places lie nearer to the Pole than any lands, which civilised man has hitherto explored. Their real habitat no human being perhaps has yet trod.

We have already mentioned the Little Stint (*Tringa minuta*), which is somewhat less polar in its habits than the preceding species, and nests moreover so far towards the south as along the Arctic Ocean coast of Asia and Europe. The same remark applies to the Grey Plover (*Squatarola helvetica*), which in Jæderen, —where it occurs like the above named, during the migration seasons,—has obtained the unmerited name of "Spanish Plover." This species nests, like the Little Stint, in the Tundra districts of North Siberia, and eastwards to the mouth of the Petchora. The Grey Phalarope *(Phalaropus fulicarius)* does not nest so far to the south as Norway.

Among the web-footed birds, besides the Brent Goose (*Berniela brenta*), already alluded to above, we are also visited by Steller's Duck (*Heniconetta stelleri*) the most prettily marked of all our ducks. This species, the male of which bears an extraordinarily variegated plumage with a silky gloss on it, never breeds west of the Murman coast,* but visits Varanger Fjord every winter in large numbers. In the summer also an occasional flock of young birds ranges about in the Fjords of Finmarken. The King Eider (*Somateria spectabilis*) moves likewise in the winter in flocks to the Finmarken coasts from its still more northerly breeding stations.

A couple of other north-easterly forms, which never nest in Norway, are the Bewick's Swan (Norwegian, Dwarf Swan, *Cygnus bewicki*), and the Smew (Norwegian, Dwarf Fish-Duck, *Mergus albellus*), and this applies probably also to the White-fronted Goose (*Anser albifrons*). As mentioned above however, it is proved, that the Barnacle Goose (*Berniela leucopsis*) may, in exceptional cases, remain to breed.

We may also mention, as non-breeding members of the gull family,—besides the white-winged gulls previously named (*Larus glaucus, L. leucopterus and Pagophila eburnea*),—the "Broad-Tailed"

* Russian Lapland; the N.W. corner of Russia, between Norway and the White Sea.— *Transl.*

(= Pomatorhine) Skua (*Stercorarius pomatorhinus*), and the Fulmar Petrel (*Fulmarus glacialis*); and of the auk family, the Spitzbergen Guillemot (*Lomvia bruennichi*), and the Little Auk (*Mergulus*).

All these species, whose breeding haunts lie further east or further north than the confines of Norway, visit our coast more or less commonly, chiefly in the winter, but a few of them also in summer. It is indeed a truly remarkable trait in these species belonging to the far north (especially the waders), which as breeding birds belong to the most northerly coast-lines of Europe, or even still more northerly regions, that many individuals often spend the summer on the most southerly coasts of Norway or still further to the south. All these consist of one- or two-year-old individuals, who are waiting until they attain to breeding age. Thus, there " summer " annually on the southern extremity of Norway (Listerland and Jæderen) flocks or stray individuals of *Tringa canutus* and *Tr. subarquata*, of the Sanderling (*Calidris*), and of the Grey Plover (*Squatarola helvetica*), and also of the Great Northern Diver (*Colymbus glacialis*), and others ; whilst the fully adult individuals of these species are hatching out their young in the extreme north; also of *Tringa minuta*, *Tr. temmincki*, of the rusty-red Bar-tailed Godwit (*Limosa lapponica*), and of the Spotted Redshank (*Totanus fuscus*), which each as breeding birds with us, belong only to Finmarken, while the young birds are to be met with in the south, all through the summer. It is as if the desire to revisit the regions in which they first saw the light, does not awake in earnest before they themselves breed.

Inside of T—— island lie a multitude of large and small holms or islands, some low, others precipitous and mountainous, and as a rule only inhabited by a pair of miserable sheep which are lodged here in the short summer time. But on most of these desolate-looking holms it is worth while for the ornithologist to go ashore. If the ground is heather-covered and swampy, there will never be lacking some nesting pairs of different species of *Tringa ;* and on the drier places, where the brownish carpet of Crowberry plants* forms comparatively large levels, interrupted by low hillocks, there will always be found established a pair of Richardson's Skuas (*Stercorarius crepidatus*), which are very skilful in hiding their dark-brown eggs or the sooty-black young, numbering one or two, among the ling. Although a near relation of the gulls, and thus like them having to seek their living on the sea, they by no means disdain anything as food, which the dry land can offer them, from Lemmings, Shrew-mice, and young birds, down to insects and Crow-berries, which they pick up among the ling.

If we approach the Skuas' nesting place, we see the two parents, the one white-, the other black-bellied, or both white-bellied, or both black, flying restlessly and silently about the spot ; now and then they throw themselves to the ground, as if struck lame, and remain lying there with extended or half-flapping wings, like birds winged by a shot, in order if possible to divert attention from their eggs or young. If we come quite close to the nest they become bolder, and at last swoop down so near us, that the tip of their wing sometimes touches one's head so forcibly that our hat flies off, and our ears tingle.

But the Skua pursues its proper craft, when it throws itself in among a flock of fishing Terns, selects an individual amongst them which has just caught a young Coal-fish of an inch long, or a Sand-Launce (*Ammodytes lancea*), and pursues it with stoops as swift as lightning until it is compelled to let go its prey, which is picked up from the surface of the water, or even while still in the air.

Although in the down all are black, the young of these variously-coloured parents are, later on, likewise varied. From similar parents (both dark-, or both white- bellied) spring similar young ones ; from variously-coloured parents spring a mixed brood, some white-bellied, some black-bellied. In the arctic parts of Norway the white-bellied individuals are the more common, on the southern coasts of the country the black.

Others of the lower holms may be entirely occupied by colonies of the Arctic Tern (*Sterna macrura*) ; on one such holm

---

* *Empetrum nigrum.*

by L——, in Laxe Fjord, thousands of pairs thus nest together, and several thousand of the small and prettily pencilled eggs, which are considered as the most savoury of all the sea birds' eggs, can generally be taken annually by the proprietor, but in certain years the whole colony absent themselves.

A phenomenon in arctic nature may yet be mentioned, which we often meet with, whilst we row about in the inner parts of the large fjords of Finmarken. From time to time we may hear far away an odd noise from the level of the sea, which sounds almost like a distant thunder clap; it is the sound produced when flocks of thousands of Eider Ducks or Velvet Scoters (*Œdemia fusca*) suddenly, as if impelled by a common inspiration, rise from the surface, fly a short distance, and settle again. These are the n n-breeding individuals, which pass the summer here in a social life devoid of cares, and perform these flying and diving exercises as a kind of game.

———

We leave T—— and the other holms in the fjord, and continue our investigations in one of the luxuriant basins which debouch upon both sides of the bottom of the fjord.* Though each of these harbours a very rich and varied fauna, we select one among them, M——, a tributary of a large river, whose broad valley bottom, situated in about 70° N. Lat., is clothed with the most luxuriant birch-woods, and whose many windings, ere it falls into the mighty head-river, form a series of flat, wooded promontories or peninsulas, here and there filled with small swamps and willow thickets. Besides these favourable localities being inhabited by their proper arctic forms among small birds and waders, several of our more southern species have also been induced to send their furthest advanced guards up hither, where they find in the sheltered sides of the valley of M——, the northernmost limit for their extension with us (and in the whole world).

We have not wandered far from the house by the river's mouth

---

* Such are the estuaries of Börs-Elv, Staburs-Elv, and Lax-Elv in Porsanger-fjord, Laxefjordbund in Laxefjord, Tana-Elv and its many tributaries inside Tana-fjord, also Nyborg and the Pasvig-Elv in Syd-Varanger.

where we lodge with the friendly Kvæn (=Finlander), Johan Kolström, before we pause at the sound of the characteristic call-note of the Pine Grosbeak (*Pinicola enucleator*), one of the prettiest birds of our fauna ; and in the (still in the beginning of July) leafless birch-trees we quickly discover its loosely built nest, composed of dry birch-twigs. Its life-history and habits are remarkable. Undismayed in the presence of danger, the Pine Grosbeak sits on its eggs so zealously, that it sometimes allows itself to be touched by hand before leaving the nest; it then flutters some few paces away, and gazes without shyness upon the strange disturber.

It is not an easy thing to give fixed rules for the variation of colour, which appears in the different ages and sexes both in this species, and in its relations the Crossbills (genus *Loxia*). In the flocks of Pine Grossbeaks, which make their appearance in the autumn on the lowlands in southern Norway, some of the males are crimson of various shades, others are yellowish ; and some greenish-yellow males may be so absolutely like the females, which are always so coloured, that no external difference can be detected between them. But in some years one meets with hardly any but red males, in others mainly yellow specimens, and it seems almost as if the young males' first plumage was in some years formed with the majority red, and in others yellow.

The commonest species of Tit in these high-lying birch-woods, is the Lapland Tit (*Parus cinctus*), which in its life-history and habits approaches the Northern Marsh Tit (*Parus borealis*), and like it hollows out its nesting-hole in the dry birch stumps. It is the "Talgoxe"* of the inhabitants of Finmarken, and comes in the winter into the houses, in order to peck at the fat in the joints of reindeer meat hanging outside the store-houses.

Before we have proceeded far along the bank of the river, where the birch wood grows luxuriantly on the warm slopes, our attention will be arrested by the voice of an un-known songster, which, with short intervals, repeats his stanza with incredible perseverance. This song, which may be heard by the hour at a time, and at all hours of the day or night, but

* Lit. Tallow-ox.—*Transl.*

especially in the warm sunshine, belongs to the Siberian Willow-Wren (*Phylloscopus borealis*). Originally known in Europe through a solitary individual, shot during the autumnal migration at Heligoland, in 1854, it was later on found several times in North Russia, but was unknown to the west of Archangel until the year 1876, when the author met with it resident at Staburnæs, in Porsanger. Its proper home is the whole of North Siberia, where its distribution extends to the east as far as Bering's Straits and Alaska, consequently through about 180° of longitude. In Finmarken it is a recent immigrant, and its migration therefore does not pass southwards along the Baltic provinces, like that of our other arctic small birds, but it migrates across the large river basins of Siberia in order to reach down to the Pacific coast, China and India, where its chief winter home is. If one could suppose that there dwelt a yearning for its native land in this little bird's breast, then an individual which picked its food, at the New Year, in the bamboo thickets by the Chinese villages or at Malacca, might recollect how it, some months previously, tried its wings for the first time in the birch-covered slopes close by North Cape, and there it would return when the spring calls it.

The song is monotonous, as with all the Willow-Wrens, and consists only of a single note, zi-zi-zi, which is repeated quickly a dozen times over; then follows a short stop, which in the height of the singing time lasts half a minute, after which follows the same strain anew, and so *ad infinitum*. This fervid song, unlike that of all other European songbirds, acts electrically upon the ornithologist, who knows that the life-history of the performer during the nesting season has hitherto been unknown to naturalists.

Until quite late years only one nest of this species had been found, namely, by Mr. Seebohm at Egasca (by the Jenisej). in July, 1877. That nest contained eggs; in July, 1885, the author found at M—— three nests, all with six or seven young.

These latter nests were placed at the foot of a tree stem, or by a tree root, where the forest was thickest, and were well concealed by the flowering *Cornus suecica*, *Veronica longifolia*, geraniums, and *Melica nutans*. As with the other Willow-Wrens they were round as a ball, provided with a roof, and loosely composed of fine grass bents, but without feathers or hair.*

The Siberian Willow-Wren with us, generally inhabits ground which is especially favourable for the development of the summer plague of Finmarken—the mosquitoes. From these they procure their food, and several times while the author has stood still to discover the secrets of the habits of this characteristic species by watching, it has through the mosquitoes been rendered almost an impossibility.

In the birch woods of M—— we meet, however, not only with species of purely arctic origin. Several genuine birds of passage, belonging to the common European fauna, also have here their furthest boundary towards the north, and the well-known notes of the Garden Warbler and the Blackcap (*Sylvia hortensis* and *S. atricapilla*), the Redstart (*Ruticilla phœnicurus*), the two Fly-catchers (*Muscicapa atricapilla* and *M. grisola*), the Tree Pipit (*Anthus trivialis*), the Song Thrush (*Turdus musicus*), the Hedge Sparrow (*Accentor modularis*), and several others, meet us frequently. They are the same forms here as in the south; the song, however, is not quite the same; it has a somewhat muffled sound, and even a few of the strains are partly different and unknown. And this allows of an easy explanation. Up here, where the areas are so large and the total number of spots habitable by these species few and far between, individuals frequently have no opportunity of hearing another of his own kind; each male sings only for his own mate, and competition can never arise, since it but seldom hears one of the same species as itself. Each one evolves his song independently of any influence from others, and they thus, therefore, acquire their individual impress.

Of other acquaintances from the lowlands, which meet us in the Finmarken valley bottoms, we may mention the House Martin (*Chelidon urbica*). In these desolate regions, this bird also knows how to adapt itself to its surroundings. As in Finmarken they only rarely find nesting-places on the few and low houses, they therefore breed in colonies, several hundred pairs together, on

---

* One of these broods is to be seen exhibited at our Zoological Museum (in Christiania), all the young ones sitting by the side of each other on a branch, the only young of that species that are as yet known to be exhibited.

the steep precipitous mountain walls, which here and there fall sheer down to the valley bottom ; here they glue up their inaccessible nests under small projections, exactly like several of the tropical species of swallow, which normally occupy cliff regions, and never come near human habitations.

The Sand Martin also (*Cotile riparia*) goes so far up to the north, as it finds serviceable nesting-places in the sand escarpments ; and wanting these, it pierces its horizontal passages into the turf roofs of the houses. The Swallow (*Hirundo rustica*) on the contrary, stops further to the south, and only one or two heedless individuals show themselves in Finmarken.

In addition, we may mention as purely accidental stragglers, only quite by chance found here in the far north, the Swift (*Cypselus apus*),* the Hoopoe (*Upupa epops*), the Turtle Dove (*Turtur communis*), the Rook (*Corvus frugilegus*), the Coot (*Fulica atra*), various waders and some other birds.

We have hinted above how the fauna in our part of the world shows a tendency to extend itself westwards, and that Finmarken has thus by degrees become inhabited by several species of purely eastern-arctic origin. It is worth remarking therefore, that there are still found various species of small birds and waders which inhabit the country to the east of Finmarken, (the Kola peninsula and north Russia), but which are not known to have yet passed the frontiers of Norway. Such are two species of Bunting (*Emberiza rustica*† and *E. pusilla*), the Yellow Wagtail with the lemon-coloured head (*Motacilla citreola*), a Pipit (*Anthus gustavi*), and the Terek Sandpiper (*Terekia terek*) ; of the Rosy Bullfinch (*Carpodacus erythrinus*) only a solitary individual blown out of his course, has hitherto been found near Christiania. The White-winged Crossbill (*Loxia leucoptera*), which nests in north Russia, is on the contrary frequently found during the autumnal

---

* It may be worth recording that in September, 1888, I saw three pairs of Swifts, which were apparently breeding on a high cliff forming the bank of the river Pasvig, half-a-day's journey eastwards of Lake Inari, in Finland, within a dozen miles or so of both the Russian and Norwegian frontiers, in approximately 68° 53′ N. Lat., and 25° 53′ E. Long.—*Transl.*

† I shot two examples on the Finnish bank of the river Tana,—that is, immediately on the Norwegian frontier,—at a place called Bildam (about midway between Polmak and Utsjok), on September 18th, 1885.—*Transl.*

and winter migration in southern Norway, but has as yet never been discovered nesting with us.

Finally it may be mentioned, that in the summer of 1876, Fisheries-Inspector Landmark found resident by Börs-Elv (in Porsanger) a Warbler, which probably was the Asiatic *Acrocephalus dumetorum*; it occurs now near Archangel, and has thus perhaps already reached our territory.

————

It remains to point out which species of birds are stationary here, and are capable of enduring the long winter with the rigorous weather and the short day-light, which prevails in these northern latitudes.

Of such species,—the hardiest outposts of the bird-world on the European continent,—we may name, among the Tits the Lapland Tit and the Northern Marsh-Tit; among the Finch tribe the Snow Bunting, the Mealy Redpoll (*Linota linaria*), and the Tree-sparrow; also, the Water-Ouzel; and of the Crow tribe the Raven, the Magpie, the Hooded Crow, and the Siberian Jay; lastly, among the Woodpeckers the Lesser Spotted Woodpecker, and the Three-toed Woodpecker.

Of gallinaceous birds, besides the two Ryper (Ptarmigan and Willow Grouse), there is also the Capercaillie, but the Black Grouse is wanting in the further districts of Finmarken, and the Hazel Grouse does not pass the middle of Nordland (Rauen).

Among the waders scarcely any other species are normally stationary than the Purple Sandpiper (*Tringa striata*); of raptorial birds the White-tailed Eagle, and the Osprey, the Gyrfalcon, and a couple of Owls (the Snowy Owl and the Hawk Owl, *Surnia ulula*), together with one or two individuals of our two common species of hawk, viz.: the Sparrow-Hawk and Goshawk; lastly, among the web-footed birds most of the Gulls, the Eider-duck, the two species of Cormorant, as well as the species of the Auk-tribe which have been already mentioned.

Some of these winter-residents among the small birds we shall touch on very briefly. In most of the fir woods, which in the larger Fjords cover the shores or the sides of the valleys, we shall frequently stumble upon a flock of Siberian Jays (*Perisoreus infaustus*), which with extended wings float about from tree to tree, and at one moment hunt for insects among the lower branches overgrown with beard-lichen,* and at another moment for berries on the ground. Everything eatable is good, and as variable as its diet is the voice, which at one moment has clear flute-like notes, and at the next is harsh, like that of its relation, the Common Jay; a curious bird, which in its life-history and habits nearly resembles a gigantic rusty-red Tit-mouse, equally inquisitive, but with something mysterious about it, and vanishing as it came, suddenly and noiselessly.

A well-known winter-visitor in our lowlands, also has here, in the northern-most conifer forests, its home and its nesting place. This is the Waxwing (*Ampelis garrulus*); vagabond and undecided in its habits as it always is, it may in certain years be found resident at various points in the interior of Finmarken, whilst in other years it is entirely invisible. For its nesting places it selects the most desolate tracts, especially where willow scrub occurs interspersed in the conifer forests; but the whole summer through it is silent and difficult to discover, which explains why an acquaintance with its breeding habits has been so comparatively lately acquired. They do not wait long after the young are fledged, but retire in flocks towards the south, only a few remaining behind to spend the winter in their native land.

As an immigrant from the south of late years, must be mentioned the Common Sparrow, which has now reached up as far as Öxfjord, to the south of Hammerfest, whilst it otherwise seems to be absent from Finmarken. Its nearest relation, the Tree-Sparrow (*Passer montanus*), which strikingly resembles the Common Sparrow, has on the contrary a wider extension, and is, where it settles itself, generally confounded with it. In 1885 I found it even established in the imposing walls, which protect the most northern fortress in the world.†

The Dipper also (*Cinclus aquaticus*) occurs by all the small swift-running rivers which are not frozen up in the winter, right up to the Arctic Ocean. The northern race of this bird is scarcely distinguishable by any constant marks from those which inhabit the water-shed of central Germany, or the mountain brooks of the Pyrenees; the extent of the brown belt

* *Usnea barbata.* † Viz. : Vardöhus.— *Transl.*

between the white breast-spot and the black belly-colour, a character on which naturalists have laid stress, has proved to vary to a wide extent within each single locality.

Several species here have apparently become the subjects of climatic variation. Let us take examples of the Northern Marsh-Tit, of the Magpie, or of the Lesser Spotted Woodpecker, already mentioned, and it will be seen that they have all altered in a certain direction; they have become whiter than the individuals from more southern localities. The Magpie occurs thus in a fine race, where the white colour of the wing feathers extends very nearly to their tips, so as to approach the White-winged Magpie (*Pica leucoptera*), from East Siberia; in *Parus borealis* the back is light gray, and the abdomen snow-white, whilst the same species down by Christiania has a darker back, and a dirty-coloured abdomen; the little Lesser Spotted Woodpecker (*Dendrocopus minor*) has a white back without the transverse bars, and nearly unspotted outer tail-feathers, answering perfectly to the east arctic *D. pipra*, whilst coexistent with them, certain individuals are dark coloured, like the normal southern stock. And the Three-toed Woodpecker has here become of a robust form, with a winter plumage of fluffy and particularly purely-coloured feathers, which are decidedly lighter than in individuals further south.

These phenomena are by no means devoid of physiological interest. It would appear that a living being is to a certain degree like a photographer's plate, which receives the impression from its surroundings. The month-long daylight in the summer and the long winter, here strive, generation after generation, to fashion individuals whiter, and have, over the whole of Norway, already given the two Ryper, the Hare, the "Snow-mouse,"* the Stoat, and the Arctic Fox, their white winter pelt, exactly as the intense sun of the tropics and the variegated splendour of colour in the vegetation there, produce the parti-coloured and metallic-glistening birds and insects, or the yellow sand of the desert is reflected in the hairy coat of the Fennec Fox† and the Jerboa, or in the feathers of the Sand Grouse (*Pteroclidæ*), and the Desert Lark.‡

---

* The so-called *Mustela nivalis* — *M. vulgaris*.—*Transl.*

† *Canis cerdo*. Gm.　　‡ *Ammonanes deserti.*

III.

FROM the confined valley bottoms with their luxuriant growth of forest, we mount up along the river-rapids; the slopes form in many places long terraces, one after another, remains of the dammings up of the water while ice-bound, monotonous in their appearance, and sterile in their nature. The forest becomes rapidly thinner, and is succeeded by scrub, and we shortly stand up on the wide plateau, which with inconsiderable interruptions of mountains or valleys, stretches over large portions of the interior of Finmarken. Upon these large moors, upon "Lapland" proper, the solitary Laplanders drive their herds of Reindeer from tract to tract, to seek the places where the Reindeers' chief food and staff of life—the Reindeer Moss (*Cladonia rangiferina*)—grows most luxuriantly; but of fixed residents there are found in these regions but few.

Innumerable lakes, most frequently surrounded by extensive stretches of bog, which are covered with dwarf-birch or willow-scrub, lie spread out upon the plateau, and are the true summer home of many of our northern species of ducks and waders.

By many of the small brooks, which purl out into the large rivers or the lakes, we shall thus find established the pretty Lesser White-fronted Goose (*Anser erythropus*), which hatches out its five grayish-yellow young under a willow-bush in the marsh adjoining one of the lakes. This small arctic species, hardly larger than a domestic duck, is widely distributed in the interior of Finmarken, but is nowhere really numerous. It seems to prefer streams with muddy banks, and thick, clay-stained water, especially if they are surrounded by thick scrub; here the Laps hunt them with dogs in the moulting season, and bring home the whole family, including the young which are as yet unable to fly, to the houses in the valley, where they soon become tame, and in the autumn have to lay down their lives like other domestic fowls.

Here and there we may also meet with a pair or two of Whoopers (*Cygnus musicus*), which build their gigantic nest, resembling an enormous ant-hill, and compounded of earth, twigs and moss, upon the small islands in the rivers or larger lakes, far from human habitations. Of other birds of the Duck family may be mentioned the Long-tailed Duck (*Harelda glacialis*), which is frequent in the small lakes in the interior of Finmarken : whilst the duck incubates her dark olive-green eggs, which are buried deep in a wreath of her brownish-black down, the drake in his variegated dress with the long swallow-tail, remains close by, and keeps a sharp look out on the trout-fry. The Goldeneye (*Clangula glaucion*) forms in several places a source of revenue for the Laplanders, who hang up for them large nest-boxes in the trees by the river banks, and in these the duck lays her eggs, which with the down are later on taken away. Naturally it lays its eggs in a hollow tree, a remarkable nesting-place for a duck ; but the way in which the newly-hatched young are slipped down to the ground by the parents in order to reach the water, is as yet scarcely explained, since this transport takes place by preference without witnesses.

If we wander about for some time on the plateaux themselves, where large tracts are covered with stones, lichens, and an extremely sparse vegetable growth, mainly consisting of *Diapensia*, Saxifrages, some graminaceous plants, and species of *Carex*, we shall soon find out how poor, considering all things, animal life is here. Hour after hour may elapse, during which we meet with scarcely a single bird. And yet certain species have their real home even in this desolate region. Thus here nests the Dotterel (*Eudromias morinellus*), and it is a cheering sight to find one of the parents, as a rule the male, leading about his small, velvety, down-covered young, pied gray and black, among the lichens and the rough grass knolls. With us this species is completely alpine in its habits, and seldom nests below the highest tree line.

A peculiar bird, half raptorial, half web-footed, also inhabits the mountain plateaux, especially where they alternate with marshes, and draws attention to itself by its odd habits. This is the Buffon's Skua (*Stercorarius parasiticus*), which unlike its relative in the belt of islands girding the coast, only makes its appearance in a single plumage with white abdomen, a black-bellied phase being unknown. With a curious cry it approaches the hunter, flaps about him in large circles, now and again remains motionless in the air, exactly like a Kestrel, and at last settles down upon a tussock to wait until the danger has passed away. Its food is as varied as that of its relative, and especially in the years in which the Lemmings undertake their mass wanderings, it lives largely upon them and various species of *Arvicola*, whilst it ordinarily persecutes young birds, and in times of scarcity contents itself with berries and insects.

Though animal life under ordinary circumstances seems to be sparse upon these extensive wastes, it is otherwise in the years in which, as above mentioned, the Lemming (*Myodes lemmus*) multiplies beyond its normal number, and undertakes its emigrations. This renowned little rodent, with its handsome yellow and black pied skin and its hasty temper, lives as a rule a little-noticed life among the tussocks and the willow-scrub upon the high mountains, and as an essentially nocturnal animal one sees under ordinary circumstances little or nothing of it.

But in certain years, under conditions which are inexplicable to us, there comes to pass their multiplication in an inordinate degree, for their prolificness during these years is almost incredible. Litter follows litter the whole summer through; young ones, which first saw the light in the spring, are already breeding in the autumn, and the old individuals often produce a new litter of young in the same nest, which the half-grown young of the preceding litter have not yet left. Gradually, as they over-run their native place, they set out upon their wanderings over the sides of the mountain; and in these great "Lemming-years" they are therefore common everywhere, alike upon the mountains and in the valleys, and by degrees reach far down upon the lowlands, where they are otherwise entirely unknown. They are now in motion at all hours of the day, they cross rivers

and lakes, bring forth young frequently on the way, are killed
during the course of the summer in thousands by men, dogs, cats
and all kinds of predaceous animals and birds, die in masses of an
epidemic, which always makes its appearance, where the condition of a species of animal is disproportionately large, and by the
arrival of autumn the hosts are already considerably diminished
in number. In the course of the winter most of them die, and
in the second year after the beginning of the emigration, there is
seldom a living Lemming to be found remaining in the valley
bottom. Not one individual returns alive to the high Fjelds.

But it is not only this species which has a mass increase in
these years. Simultaneously other small rodents, especially
of the genus *Arvicola*, increase beyond the normal number ;
thus in Finmarken the Gray-sided Field-Mouse (*A. rufocanus*),
upon the high Fjelds the Mountain Rat (*A. ratticeps*), in the
forest districts the Forest Lemming (*Myodes schisticolor*), and the
Bank Vole (*A. glareolus*), or upon the lowlands the common *A.
agrestis ;* yes, even the Shrews, the Hares, and the Ryper are
wont to be more numerous in these years than in others.

Then come also all kinds of raptorial birds and carnivorous
animals, springing as it were from the ground, enticed to the
place by the profusion of food. The mountains swarm with
various hawks, especially Rough-legged Buzzards (*Archibuteo
lagopus*) and various species of owl ; especially Short-eared Owls
(*Asio brachyotus*) and Snowy Owls ; and Gyr-falcons (*Hierofalco
gyrfalco*) now show themselves comparatively frequently. Among
the predaceous animals are Arctic Foxes, Stoats, and Weasels,
ubiquitous everywhere ; and they all live almost exclusively on
the Lemmings. Even the fish are among their enemies ; one
may, not unfrequently in such a year, come across a trout with
its belly distended by a Lemming, which the voracious fish has
swallowed, while it was essaying to cross a river or lake.

Even among the raptorial birds the prolificness of the year
leaves its traces. The Snowy Owl (*Nyctea scandiaca*), which inhabits all the mountain plateaux of the country right up to North
Cape (and Spitzbergen), but which in ordinary years is found
scattered and sparingly, is, during these breeding years, so
numerous as to be scarcely absent from any part of the mountain,

and at the same time so prolific that they have as many as ten eggs in their nest.

The process of hatching of such an unusual clutch for a raptorial bird is abnormal. We stand by the side of such a nest, which lies on the open mountain, generally in the neighbourhood of moist ground. In the nest lie four half-grown young of the size of a hazel grouse, with well developed wings, which in a fortnight's time will be ready to do service; in addition the nest contains two young ones, which are considerably smaller, and still half down-clad; lastly three ditto, which are just hatched, and lie half buried under their bigger brothers and sisters. Under these nine young ones we shall finally find an egg, which is perhaps only half incubated, or nearly ready to hatch. These many stages among the young arise from the fact that an interval of several days may often elapse between the laying of one or a pair of eggs and the next; so soon as the first young ones have burst the egg shell, the parents are obliged to seek food for them, and the incubation of the remaining eggs is thus given up, altogether or in part, to the elder brothers and sisters. In the incubation of the last egg at any rate the parents take hardly any part.

Whilst the hen is sitting, the somewhat smaller and nearly snow-white old cock bird sits on guard in the vicinity of the nest, and warns his mate by a strange shriek, if a hunter approaches; both then circle about him with their peculiar slow wing-strokes, and swoop impetuously straight down upon his dog, which may easily be wounded thereby. The hen however in general takes the affair more calmly; she is more wary than the male, and fatter, since she is always fed by him. Around the nest lies a rich supply of Lemmings and Rype-chicks, whole or dismembered, and always in larger quantities than the young are capable of consuming.

This successive development of the young may be traced in ordinary years also, in other raptorial birds as well, but is most perceptible during these prolific years, since the clutch of eggs is often, _e.g._, with the Rough-legged Buzzard, or with the Hawk-Owl (_Surnia funerea_), increased beyond the normal number.

Our excursions on the moors are nearly concluded. We are on the point of setting out on our return journey, tired of the boggy hollows, the willow-scrubs, or the interminable tracts covered with loose stones; but retaining the impression of a remarkable, and in many respects grand, nature. A peculiarly harsh bird-cry rings about us from a group of birch-trees, which grow here and there on the flat lichen-covered terrace that forms the last step before coming to the valley bottom. We soon see that it proceeds from a pair of Great Grey Shrikes (*Lanius excubitor*), who are apprehensive about their nest; this lies large and exposed in the top of the biggest of the birches, well-lined with white Rype feathers, which at a long distance can be seen protruding from the side of the nest—an enticing sight for an ornithologist. Here he will be able to obtain a contribution towards the solution of a question which has often been propounded —which race of this species of bird it is which inhabits arctic Europe, whether it is the "single-spotted" form, whose wing only has one large white spot by the root of the quill feathers ("*L. major*"), or the typical one, which, besides this spot on the primaries, has also a large white spot by the root of the secondaries, and therefore may be called the "Two-spotted." Both are found indiscriminately during the migration seasons and in winter, in nearly all other parts of Europe, and several naturalists have therefore believed that the single-spotted and somewhat darker "*Lanius major*" was a really arctic species, which had its home in northern Europe and northern Asia, where the typical *Lanius excubitor* would not be found.

Within five minutes both examples lie in our hand; but the solution of the question is apparently just as far off as before. For the fact is, the male proves to be a typical *L. excubitor*, the female an equally strongly-marked "*L. major*," and the young in the nest are too small to afford any evidence.

We learn hence, however, this much, that both forms are only varieties, which may occur indiscriminately in these northern regions; but they are not good species, since they pair with each other (which two distinct species normally never do). The typical European form is unquestionably the two spotted *L. excubitor*; but it has a strong tendency to variation in the distribution of the white colour ; these light and dark varieties are often found paired with each other, and producing individuals which partly resemble both parents, partly constitute intermediate forms between them.

We have completed the tour which we had proposed to ourselves. We have picked up some of the scattered features of the life-history of a single group of animals in the arctic parts of our country, as they occur to us among the numerous islands girding the coast, on the birch-covered slopes in the sheltered valleys, and upon the treeless expanses of the mountains. Our porters stand ready, outside the little mountain hut, which for several days has afforded us a roof over our heads, and we retire again into the valley, where the first trace of civilisation begins to show itself.

But we only leave these regions reluctantly, with their marvellously attractive nature, thin pure air, and the free, unconstrained life. And as we stand on deck, and see the last snow-covered peaks which defend the Land of the Midnight Sun disappear under the horizon, and return again to " the gilded misery " (as the enthusiastic Englishmen say when they take leave of our mountains), a yearning draws us strongly back towards those regions which are apparently so poor, but which nevertheless yield to every lover of nature, memories never-to-be-forgotten.

APPENDIX

APPENDIX.

A List of the Birds of Norway, arranged according to the Rules of the B.O.U.

The total number of Birds, hitherto found in Norway, is	278

The number of Breeding Species is :—

Regular (normal) Breeders	190
Rare Breeders	7
Uncertain Breeders	15
	212

The number of Non-breeding Visitors, is :—

Annual (normal) Visitors	12
Rare Visitors	13
" Meteoric " Visitor	1
Accidental Visitors	40
	66

Besides these there are 3 Hybrids.

A.—BREEDING SPECIES.

1. Regular (Normal) Breeders (190 Species).

Family TURDIDÆ

Subfamily TURDINÆ

Missel-Thrush. *Turdus viscivorus*, Linn.
Song-Thrush. *Turdus musicus*, Linn.
Redwing. *Turdus iliacus*, Linn.
Fieldfare. *Turdus pilaris*, Linn.
Blackbird. *Turdus merula*, Linn.
Ring-Ouzel. *Turdus torquatus*, Linn.
Wheatear. *Saxicola œnanthe*, (Linn.)
Whinchat. *Pratincola rubetra*, (Linn.)
Redstart. *Ruticilla phœnicurus* (Linn.)
Red-spotted Bluethroat. *Cyanecula suecica*, (Linn.)
Redbreast. *Erithacus rubecula*, (Linn.)

Subfamily SYLVIINÆ

Whitethroat. *Sylvia cinerea*, Bechst.
Lesser Whitethroat. *Sylvia curruca*, (Linn.)
Blackcap. *Sylvia atricapilla*, (Linn.)

Subfamily SYLVIINÆ — *continued*

Garden Warbler. *Sylvia hortensis*, Bechst.
Goldcrest. *Regulus cristatus*, Koch.
Chiffchaff. *Phylloscopus rufus*, (Bechst.)
Willow-Warbler. *Phylloscopus trochilus*, (Linn.)
Siberian Willow-Warbler. *Phylloscopus borealis*, (Blasius.)
Icterine Warbler. *Hypolais icterina*, (Vieill.)
Sedge-Warbler. *Acrocephalus phragmitis*, (Bechst.)

Subfamily ACCENTORINÆ

Hedge-sparrow. *Accentor modularis*, (Linn.)

Family CINCLIDÆ

Dipper. *Cinclus aquaticus*, Bechst.

A.—BREEDING SPECIES.

1. Regular (Normal) Breeders (190 Species)—*continued*

Family PARIDÆ

White-headed Long-tailed Titmouse. *Acredula caudata*, (Linn.)
Great Titmouse. *Parus major*, Linn.
Continental Coal Titmouse. *Parus ater*, Linn.
Marsh-Titmouse. *Parus palustris*, Linn.
Northern Marsh-Tit. *Parus borealis*, De Selys.
Lapland Tit. *Parus cinctus*, Bodd.
Blue Titmouse. *Parus cœruleus*, Linn.
Crested Titmouse. *Parus cristatus*, Linn.

Family SITTIDÆ

Northern Nuthatch. *Sitta europæa*, Linn.

Family TROGLODYTIDÆ

Wren. *Troglodytes parvulus*, Koch.

Family MOTACILLIDÆ

White Wagtail. *Motacilla alba*, Linn.
Blue-headed Yellow Wagtail. *Motacilla flava*, Linn.
Meadow-Pipit. *Anthus pratensis*, (Linn.)
Red-throated Pipit. *Anthus cervinus*, (Pall.)
Tree-Pipit. *Anthus trivialis*, (Linn.)
Rock-Pipit. *Anthus obscurus*, (Lath.)

Family LANIIDÆ

Great Grey Shrike. *Lanius excubitor*, Linn.
Red-backed Shrike. *Lanius collurio*, Linn.

Family AMPELIDÆ

Waxwing. *Ampelis garrulus*, Linn.

Family MUSCICAPIDÆ

Spotted Flycatcher. *Muscicapa grisola*, Linn.
Pied Flycatcher. *Muscicapa atricapilla*, Linn.

Family HIRUNDINIDÆ

Swallow. *Hirundo rustica*, Linn.
Martin. *Chelidon urbica*, (Linn.)
Sand-Martin. *Cotile riparia*, (Linn.)

Family CERTHIIDÆ

Tree-creeper. *Certhia familiaris*, Linn.

Family FRINGILLIDÆ

Goldfinch. *Carduelis elegans*, Steph.
Siskin. *Chrysomitris spinus*, Gould.
Greenfinch. *Ligurinus chloris*, Gould.
House-Sparrow. *Passer domesticus*, (Linn.)
Tree-Sparrow. *Passer montanus*, (Linn.)
Chaffinch. *Fringilla cœlebs*, Linn.
Brambling. *Fringilla montifringilla*, Linn.
Linnet. *Linota cannabina*, (Linn.)
Mealy Redpoll. *Linota linaria*, (Linn.)
Twite. *Linota flavirostris*, (Linn.)

Subfamily LOXIINÆ

Northern Bullfinch. *Pyrrhula major*, Brehm.
Pine Grosbeak. *Pinicola enucleator*, (Linn.)
Parrot Crossbill. *Loxia pityopsittacus*, Bechst.
Crossbill. *Loxia curvirostra*, Linn.

Subfamily EMBERIZINÆ

Corn-Bunting. *Emberiza miliaria*, Linn.
Yellow Hammer. *Emberiza citrinella*, Linn.
Ortolan Bunting. *Emberiza hortulana*, Linn.
Reed-Bunting. *Emberiza schœniclus*, Linn.
Lapland Bunting. *Calcarius lapponicus*, (Linn.)
Snow-Bunting. *Plectrophanes nivalis*, (Linn.)

Family STURNIDÆ

Starling. *Sturnus vulgaris*, Linn.

Family CORVIDÆ

Nutcracker. *Nucifraga caryocatactes*, (Linn.)
Siberian Jay. *Perisoreus infaustus*, (Linn.)
Jay. *Garrulus glandarius*, (Linn.)
Magpie. *Pica rustica*. (Scop.)
Jackdaw. *Corvus monedula*, Linn.
Hooded Crow. *Corvus corniæ*, Linn.
Rook. *Corvus frugilegus*, Linn.
Raven. *Corvus corax*, Linn.

A.—BREEDING SPECIES.

1. Regular (Normal) Breeders (190 Species)—continued

Family ALAUDIDÆ

Sky-Lark. *Alauda arvensis*, Linn.
Wood-Lark. *Alauda arborea*, Linn.
Shore-Lark. *Otocorys alpestris*, (Linn.)

Family CYPSELIDÆ

Swift. *Cypselus apus*, (Linn.)

Fam. CAPRIMULGIDÆ

Nightjar. *Caprimulgus europœus*, Linn.

Family PICIDÆ

Subfamily PICINÆ

Great Black Woodpecker. *Picus martius*, Linn.
Great Spotted Woodpecker. *Dendrocopus major*, (Linn.)
White-backed Woodpecker. *Dendrocopus leuconotus*, Bechst.
Lesser Spotted Woodpecker. *Dendrocopus minor*, (Linn.)
Three-toed Woodpecker. *Picoides tridactylus*, (Linn.)
Green Woodpecker. *Gecinus viridis*, (Linn.)
Grey-headed Green Woodpecker. *Gecinus canus*, (Gmel.)

Subfamily IYNGINÆ

Wryneck. *Iynx torquilla*, Linn.

Family CUCULIDÆ

Cuckoo. *Cuculus canorus*, Linn.

Family ASIONIDÆ

Long-eared Owl. *Asio otus*, (Linn.)
Short-eared Owl. *Asio brachyotus*, (Forster.)
Tawny Owl. *Syrnium aluco*, (Linn.)
Ural Owl. *Syrnium uralense*, (Pall.)
Lapp Owl. *Syrnium lapponicum*, (Sparrm.)
Snowy Owl. *Nyctea scandiaca*, (Linn.)
European Hawk-Owl. *Surnia ulula*, (Linn.)
Tengmalm's Owl. *Nyctala tengmalmi*, (Gmel.)
Eagle Owl. *Bubo ignavus*, Newton.
Pygmy Owl. *Glaucidium passerinum*, (Linn.)

Family FALCONIDÆ

Buzzard. *Buteo vulgaris*, Leach.
Rough-legged Buzzard. *Archibuteo lagopus*, (Gmel.)

Family FALCONIDÆ—continued

Golden Eagle. *Aquila chrysaëtus*, (Linn.)
White-tailed Eagle. *Haliaëtus albicilla*, (Linn.)
Gos-Hawk. *Astur palumbarius*, (Linn.)
Sparrow-Hawk. *Accipiter nisus*, (Linn.)
Gyr Falcon. *Hierofalco gyrfalco*, (Linn.)
Peregrine Falcon. *Falco peregrinus*, Tunstall.
Hobby. *Falco subbuteo*, Linn.
Merlin. *Falco æsalon*, Tunstall.
Kestrel. *Tinnunculus alaudarius*, (Gmel.)
Osprey. *Pandion haliaëtus*, (Linn.)

Family PELECANIDÆ

Cormorant. *Phalacrocorax carbo*, (Linn.)
Shag. *Phalacrocorax graculus*, (Linn.)

Family ARDEIDÆ

Heron. *Ardea cinerea*, Linn.

Family ANATIDÆ

Gray Lag Goose. *Anser cinereus*, Meyer.
Bean-Goose. *Anser segetum*, (Gmel.)
Lesser White-fronted Goose. *Anser erythropus*, (Linn.)
Whooper Swan. *Cygnus musicus*, Bechst.
Common Sheldrake. *Tadorna cornuta*, (Gmel.)
Wigeon. *Mareca penelope*, (Linn.)
Pintail. *Dafila acuta*, (Linn.)
Wild Duck. *Anas boscas*, Linn.
Common Teal. *Querquedula crecca*, (Linn.)
Tufted Duck. *Fuligula cristata*, (Leach.)
Scaup. *Fuligula marila*, (Linn.)
Goldeneye. *Clangula glaucion*, (Linn.)
Long-tailed Duck. *Harelda glacialis*, (Linn.)
Eider Duck. *Somateria mollissima*, (Linn.)
Common Scoter. *Œdemia nigra*, (Linn.)
Velvet Scoter. *Œdemia fusca*, (Linn.)
Goosander. *Mergus merganser*, Linn.
Red-breasted Merganser. *Mergus serrator*, Linn.

A.—BREEDING SPECIES.

1. Regular (Normal) Breeders (190 Species)—*continued*

Family COLUMBIDÆ

Ring-Dove. *Columba palumbus*, Linn.
Stock-Dove. *Columba œnas*, Linn.

Family PHASIANIDÆ

Partridge. *Perdix cinerea*, Lath.
Quail. *Coturnix communis*, Bonnat.

Family TETRAONIDÆ

Ptarmigan. *Lagopus mutus*, (Montin.)
Willow Grouse. *Lagopus albus*, (Gmel.)
Hazel Grouse. *Bonasa betulina*, (Scop.)
Black Grouse. *Tetrao tetrix*, Linn.
Capercaillie. *Tetrao urogallus*, Linn.

Family RALLIDÆ

Water-Rail. *Rallus aquaticus*, Linn.
Spotted Crake. *Porzana maruetta*, (Leach.)
Corn-Crake. *Crex pratensis*, Bechst.
Moor-hen. *Gallinula chloropus*, (Linn.)

Family GRUIDÆ

Crane. *Grus communis*, Bechst.

Family CHARADRIIDÆ

Golden Plover. *Charadrius pluvialis*, Linn.
Little Ringed Plover. *Ægialitis euronica*, (Gmel.)
Ringed Plover. *Ægialitis hiaticula*, (Linn.)
Dotterel. *Eudromias morinellus*, (Linn.)
Lapwing. *Vanellus vulgaris*, Bechst.
Turnstone. *Strepsilas interpres*. (Linn.)
Oyster-catcher. *Hæmatopus ostralegus*, Linn.

Family SCOLOPACIDÆ

Red-necked Phalarope. *Phalaropus hyperboreus*. (Linn.)
Woodcock. *Scolopax rusticula*, Linn.
Great Snipe. *Gallinago major*, (Gmel.)
Common Snipe. *Gallinago cœlestis*, (Frenzel)
Jack Snipe. *Limnocryptes gallinula*, (Linn.)
Broad-billed Sandpiper. *Limicola platyrhyncha*, (Temm.)
Dunlin. *Tringa alpina*, Linn.
Temminck's Stint. *Tringa temmincki*, Leisl.

Family SCOLOPACIDÆ—*continued*

Purple Sandpiper. *Tringa striata*, Linn.
Ruff. *Machetes pugnax*, (Linn.)
Common Sandpiper. *Tringoides hypoleucus* (Linn.)
Green Sandpiper. *Helodromas ochropus*, (Linn.)
Wood-Sandpiper. *Totanus glareola*, (Gmel.)
Redshank. *Totanus calidris*. (Linn.)
Spotted Redshank. *Totanus fuscus*, (Linn.)
Greenshank. *Totanus canescens*, (Gmel.)
Bar-tailed Godwit. *Limosa lapponica*, (Linn.)
Whimbrel. *Numenius phæopus*, (Linn.)
Curlew. *Numenius arquata*, (Linn.)

Family LARIDÆ

Subfamily STERNINÆ

Arctic Tern. *Sterna macrura*, Naum.
Common Tern. *Sterna fluviatilis*, Naum.

Subfamily LARINÆ

Kittiwake. *Rissa tridactyla*, (Linn.)
Herring-Gull. *Larus argentatus*, Gmel.
Lesser Black-backed Gull. *Larus fuscus*, Linn.
Common Gull. *Larus canus*, Linn.
Greater Black-backed Gull. *Larus marinus*, Linn.

Subfamily STERCORARIINÆ

Richardson's Skua. *Stercorarius crepidatus*, (Gmel.)
Buffon's Skua. *Stercorarius parasiticus*, (Linn.)

Family COLYMBIDÆ

Black-throated Diver. *Colymbus arcticus*, Linn.
Red-throated Diver. *Colymbus septentrionalis*, Linn.

Family PODICIPIDÆ

Sclavonian Grebe. *Podiceps auritus*, (Linn.)

Family ALCIDÆ

Razorbill. *Alca torda*, Linn.
Common Guillemot. *Lomvia troile*, (Linn.)
Black Guillemot. *Uria grylle*, (Linn.)
Puffin. *Fratercula arctica*, (Linn.)

A.—BREEDING SPECIES.

1. Regular (Normal) Breeders (190 Species)—*continued*.

Among these 190 species of "Normal Breeders," the particular breeding places of three species are still unknown, namely, of *Limnocryptes gallinula*, *Totanus fuscus*, and *Limosa lapponica*. Two species breed in so limited numbers, that they might with almost equal propriety have been classed under "Rare breeders," namely, *Emberiza miliaria* and *Corvus frugilegus*. It must also be noted, that, by *Motacilla flava* is understood both the type race (which breeds extremely sparingly in Norway) and the North European form *M. viridis*, Gmel.

---

2. Rare Breeders (7 Species).

Family TURDIDÆ
Subfamily SYLVIINÆ
Wood-Warbler. *Phylloscopus sibilatrix*, (Bechst.)

Family MOTACILLIDÆ
Pied Wagtail. *Motacilla lugubris*, Temm.

Family FALCONIDÆ
Hen-Harrier. *Circus cyaneus*, (Linn.)
Kite. *Milvus ictinus*, Savigny.

Family RALLIDÆ
Coot. *Fulica atra*, Linn.

Family CHARADRIIDÆ
Kentish Plover. *Ægialitis cantiana*, (Lath.)

Family SCOLOPACIDÆ
Little Stint. *Tringa minuta*, Leisl.

---

3. Uncertain Breeders (15 Species).

Family TURDIDÆ
Subfamily SYLVIINÆ
Grasshopper Warbler. *Locustella nævia*, (Bodd.)

Family FRINGILLIDÆ
Hawfinch. *Coccothraustes vulgaris*, Pall.

Subfamily LOXIINÆ
White-winged Crossbill. *Loxia leucoptera*, Gmel.

Family ALCEDINIDÆ
Kingfisher. *Alcedo ispida*, Linn.

Family UPUPIDÆ
Hoopoe. *Upupa epops*, Linn.

Family ANATIDÆ
Barnacle Goose. *Bernicla leucopsis*, (Bechst.)
Garganey. *Querquedula circia*, (Linn.)

Family ANATIDÆ—*continued*
Shoveller. *Spatula clypeata*, (Linn.)

Family COLUMBIDÆ
Rock-Dove. *Columba livia*, Bonat.
Turtle Dove. *Turtur communis*, Selby.

Family LARIDÆ
Subfamily LARINÆ
Black-headed Gull. *Larus ridibundus*, Linn.

Family COLYMBIDÆ
Great Northern Diver. *Colymbus glacialis*, Linn.

Family PODICIPIDÆ
Great Crested Grebe, *Podiceps cristatus* (Linn.)
Red-necked Grebe. *Podiceps griseigena*, (Bodd.)
Little Grebe. *Tachybaptes fluviatilis*, (Tunstall.)

All these "Uncertain Breeders" make their appearance only sparingly, generally during the autumnal migration, or in the winter; a few only in summer. *Locustella nævia* has only been found in the country on two occasions, on the latter of which it was probably breeding. *Columba livia* is probably now exterminated from the fauna of the country.

B.—NON-BREEDING VISITORS.

1. Annual (Normal) Visitors (12 Species).

Family PELECANIDÆ
Gannet. *Sula bassana*, (Linn.)

Family ANATIDÆ
White-fronted Goose. *Anser albifrons*, (Scop.)
Brent Goose. *Bernicla brenta*, (Pall.)
Steller's Duck. *Heniconetta stelleri*, (Pall.)
King Eider. *Somateria spectabilis*, (Linn.)

Family CHARADRIIDÆ
Grey Plover. *Squatarola helvetica*, (Linn.)

Family SCOLOPACIDÆ
Curlew Sandpiper. *Tringa subarquata*. (Gühl.)
Knot. *Tringa canutus.* Linn.
Sanderling. *Calidris arenaria.* (Linn.)

Family LARIDÆ
Subfamily LARINÆ
Glaucous Gull. *Larus glaucus*, Fabr.

Family PROCELLARIIDÆ
Fulmar. *Fulmarus glacialis*, (Linn.)

Family ALCIDÆ
Little Auk. *Mergulus alle*, (Linn.)

To these "Annual Visitors" may be added the easterly form of *Nucifraga caryocatactes* (forma *macrorhynchus*, Brehm). Two of the above named species, *Somateria spectabilis*, and *Larus glaucus*, should possibly be classed under the "Uncertain breeders."

2. Rare Visitors (12 Species).

Family CICONIIDÆ
White Stork. *Ciconia alba*, Bechst.

Family ANATIDÆ
Pink-footed Goose. *Anser brachyrhynchus*, Baill.
Bewick's Swan. *Cygnus bewicki.* Yarr.
Pochard. *Fuligula ferina* (Linn.)

Family SCOLOPACIDÆ
Grey Phalarope. *Phalaropus fulicarius*, (Linn.)

Family LARIDÆ
Subfamily STERNINÆ
Black Tern. *Hydrochelidon nigra*, (Linn).

Subfamily LARINÆ
Ivory Gull. *Pagophila eburnea*, (Phipps)
Iceland Gull. *Larus leucopterus*, Faber.

Subfamily STERCORARIINÆ
Pomatorhine Skua. *Stercorarius pomatorhinus*, (Gould)

Family PROCELLARIIDÆ
Storm-Petrel. *Procellaria pelagica*, Linn.
Manx Shearwater. *Puffinus anglorum*, (Temm.)

Family ALCIDÆ
Brünnich's Guillemot. *Lomvia bruennichi*, (Sabine)

3. Meteoric Visitor (1 Species.)

Family PTEROCLIDÆ.
Pallas's Sand-Grouse. *Syrrhaptes paradoxus*, (Pall).

B.—NON-BREEDING VISITORS.

4. Accidental Visitors (41 Species.)

Family TURDIDÆ

Subfamily TURDINÆ

Dusky Thrush. *Turdus fuscatus*, Pall.
Black-throated Thrush. *Turdus atrigularis*, Temm.
White's Thrush. *Turdus varius*, Pall.
Stonechat. *Pratincola rubicola*, (Linn.)
Black Redstart. *Ruticilla titys*, (Scop.)

Family MOTACILLIDÆ

Grey Wagtail. *Motacilla melanope*, Pall.
Yellow Wagtail. *Motacilla raii*, (Bp.)
Richard's Pipit. *Anthus richardi*, Vieill.

Family ORIOLIDÆ

Golden Oriole. *Oriolus galbula*, Linn.

Family FRINGILLIDÆ

Subfamily LOXIINÆ

Rosy Bullfinch. *Carpodacus erythrinus*, (Pall.)

Family STURNIDÆ

Rose-coloured Pastor. *Pastor roseus*, (Linn.)

Family CORVIDÆ

Carrion-Crow. *Corvus corone*, Linn.

Family ALAUDIDÆ

Crested Lark. *Alauda cristata*, Linn.

Family CORACIIDÆ

Roller. *Caracias garrula*, Linn.

Family FALCONIDÆ

Marsh-Harrier. *Circus æruginosus*, (Linn.)
Pallid Harrier. *Circus swainsoni*, Smith.
* Greenland Falcon. *Hierofalco islandus*, (Gmel.)
* Iceland Falcon. *Hierofalco rusticolus*, (Fabr.)

Family ARDEIDÆ

Purple Heron. *Ardea purpurea*, Linn.
Little Bittern. *Ardetta minuta*, (Linn.)
Bittern. *Botaurus stellaris*, (Linn.)

Family CICONIIDÆ

Black Stork. *Ciconia nigra*, (Linn.)

Family PLATALEIDÆ

Spoonbill. *Platalea leucorodia*, Linn.
Glossy Ibis. *Plegadis falcinellus* (Linn.)

Family ANATIDÆ

Snow-Goose. *Chen hyberboreus*, (Pall.)
Mute Swan. *Cygnus olor*, (Gmel.)
Ruddy Sheldrake. *Tadorna casarca*, (Linn.)
Gadwall. *Chaulelasmus streperus*, (Linn.)
Barrow's Goldeneye. *Clangula islandica*. (Gmel.)
Smew. *Mergus albellus*, Linn.

Family OTIDÆ

Little Bustard. *Otis tetrax*, Linn.

Family GLAREOLIDÆ

Nordmann's Pratincole. *Glareola melanoptera*, Nordm.

Family SCOLOPACIDÆ

Avocet. *Recurvirostra avocetta*, Linn.

Family LARIDÆ

Subfamily STERNINÆ

Sandwich Tern. *Sterna cantiaca*, Gmel.

Subfamily LARINÆ

Little Gull. *Larus minutus*, Pall.
Sabine's Gull. *Xema sabinii*, (Sabine)

* See next page.

B.-- NON-BREEDING VISITORS.

4. Accidental Visitors (41 Species)—*continued.*

Family LARIDÆ
Subfamily STERCORARIINÆ
Common Skua. *Stercorarius catarr-hactes*, (Linn.)

Family PROCELLARIIDÆ
Leach's Petrel. *Procelleria leucorrhoa*, Vieill.
Greater Shearwater. *Puffinus major*, Faber.

Family COLYMBIDÆ
Yellow-billed Diver. *Colymbus adamsi*, Gray.

Family PODICIPIDÆ
Eared Grebe. *Podiceps nigricollis*, C. L. Brehm.

Amongst the above enumerated forty-one "Accidental Visitors," are counted as distinct species, the Greenland Falcon.* *Hierofalco islandus*, and the Iceland Falcon, *Hierofalco rusticolus*,* whose title to distinct specific rank is disputed.

C.—HYBRIDS.

Family TETRAONIDÆ

Tetrao tetrix ♂ + *Tetrao urogallus* ♀ ("Rakkel-Hane," male and female).
Lagopus albus ♂ (?) + *Tetrao tetrix* ♀ (?) ("Rype-Orre," male and female).
Lagopus albus ♂ + *Tetrao urogallus* ♀ ("Rype-Tiur," male.)

---

* This nomenclature for the Gyrfalcon (or · falcons) is best explained by a quotation from a former paper by Prof. Collett, entitled "Om 6 for Norges Fauna nye Fugle, fundne in 1887-1889" (*Christa. Vidensk. Sels. Forh.*, 1890, No. 4, p. 7) —*Transl.*

F. islandus, or the white Greenland falcon, which is most frequently referred to under Gmelin's later name of *F. candicans*, is known principally from Greenland, and the majority of examples preserved in the Museums come from there. It nests there in the more northern districts (north of the Arctic circle), but is tolerably frequent in South Greenland during the periods of migration and in the winter.

In the southern parts of Greenland there also occurs, very numerously, the real Iceland Falcon, which was named by Gmelin in 1788, *F. islandus*, under which name it has hitherto been entered by most writers (including the present author in *Nyt. Mag. f. Nature.*, B. 26, p. 329). This form, which is regarded by most writers as a light climatic race of the North European *F. gyrfalco*, was as long ago as 1780 named by Fabricius in his *Fauna Groenlandica*, *F. rusticolus* (and its immature form, *F. fuscus*), after the species described in 1766 under that name in Linnæus' *Syst. Nat.*, ed. xii. Although Linnæus' description of *F. rusticolus* in the place quoted, may no doubt pass for the Iceland species, the probability is somewhat lessened that he really had it before him, because he states as its habitat: "*ex Succia*," a country, where the species can, at the present time, scarcely be said with safety to occur. It is therefore safest to date the name *F. rusticolus* for the Iceland species from Fabricius 1780, since that author unquestionably had just that species before his eyes (and likewise the white Greenland falcon for his description of *F. islandus*).

www.ingramcontent.com/pod-product-compliance
Lightning Source LLC
Chambersburg PA
CBHW030855260626
47169CB00008B/2547